# GENERAL SYSTEMS THEORY

# A focus on computer science engineering

## DOUGGLAS HURTADO CARMONA

# General Systems Theory
**A focus on computer science engineering**

**Dougglas Hurtado Carmona**

© **2011, Copyright of this edition:**
Dougglas Hurtado Carmona
ISBN: **978-1-257-78224-6**

Translated from:
**Teoría General de Sistemas: Un enfoque hacia la ingeniería de sistemas.**

© 2010. ISBN: **978-958-44-6487-3**

**More Information:**
dhurtado@samartinbaq.edu.co
dougglash@yahoo.com.mx
dougglas@gmail.com

**Front cover:**
Sander Cadena Hernández

# ACKNOWLEDGEMENTS

*GOD Almighty.*

*Fundación Universitaria San Martín, especially Dr. Joseph Santiago Alvear, and the Faculty of Engineering (Jorge, Lucho, Horatio Nelson and Karol).*

*My friends Ralph Rodriguez and Roberto Salas.*

*My wife Luisa and my chindren Dougglas David and Maauxi Andrea.*

*My first "Pupil"*
Carlos Acosta, Edwin Andrade, Carmen Barraza, María Barros, Samir Cuello, Trinidad De Alba, Jerson Eguis, Israel Escobar, Gunther Hillenbrand, Shirley Jiménez, Gustavo Julio, Deivis Lozano, Jaime Maury, Luis Carlos Mercado, Luz Milena Mora, Eduardo Muvdi, Nazly Olmos, Winsthon Peláez, Guillermo Rodríguez Mass, Luz Karime Rico, Gabriel Salas, Jean Carlos Salinas, Carlos Sánchez, Johana Santana, Lia Stand, Jorge Sugura, Juan Silva, Danilo Torres y Jorge Vengoechea.

# AUTHOR

## DOUGGLAS HURTADO CARMONA

*Master of Computer Systems Engineering, Computer Systems Engineering, Minor in Business Administration and Information Systems Security. IBM Certification in Information Systems Management. Diploma in Scientific Research, Development for Web Applications, Computer Security and Forensic Computing, and in education and pedagogy.*

Since 2002 he works as Head of Research at the Faculty of Engineering of the San Martín University Foundation Headquarters Puerto Colombia, Barranquilla - Colombia.

National and international lecturer with 12 years of university teaching experience in the areas of Programming Objects, Data Structures Object Oriented, Systems Theory, Systems Analysis and Design, Operating Systems, Compilers, Databases, Programming Concurrent and Client / Server Java, application development for Internet, computer security, computer forensics and contingency plans.

Researcher on the topics of Information Security, Computer Forensics, General Systems Theory and dynamic systems for software engineering, and compiler theory. Developed the research "Analysis of skill development from the use of Computer Assisted Learning" which received Special Mention Awards ACOFI 2007, and "Methodology for the development of systems based on learning objects." Creator OSOFFICE, Educational Software for Teaching Operating Systems.

He has served as Director of software development projects, systems analyst and programmer, IT project manager, information security engineer, independently has advised companies involved in building software.

# TABLE OF CONTENT

## 1 BASES ON THE GENERAL SYSTEMS THEORY — 1
- REDUCTIONIST APPROACH — 1
  - *Specialization* — *1*
  - *Reductionist theory* — *1*
- THE APPROACH OF GENERAL SYSTEMS THEORY — 2
  - *Approaches to general systems theory* — *2*
  - *Frameworks for the study of G.S.T.* — *3*
  - *Trends of practical application of G.S.T.* — *4*
- FOCUS ART TO SOLVE PROBLEMS — 6
  - *What is a problem?* — *6*
  - *First approach: modeling from reality* — *7*
  - *Second approach: creativity and constraints* — *8*

## 2 FOUNDATIONS OF SYSTEMS — 11
- BASIC DEFINITIONS — 11
  - *Definition of Energy* — *11*
  - *Definition of system* — *11*
  - *Definition of Mega-system* — *12*
  - *Definition of Super-System* — *12*
  - *Definition of SubSystem* — *12*
- ELEMENTS OF A SYSTEM — 13
  - *Objectives* — *13*
  - *Synergy* — *14*
  - *Recursion* — *14*
  - *The input currents - Inflows* — *15*
  - *The conversion process* — *16*
  - *The Output currents - Outflows* — *16*
  - *Feedback communication - The provision of feedback* — *17*
  - *Frontiers of the system* — *17*
  - *Environment System* — *18*
- LEVELS OF ORGANIZATION OF SYSTEMS — 18
- ENTROPY IN THE SYSTEMS — 19

| | |
|---|---:|
| *Entropy* | *19* |
| *Negative Entropy* | *20* |
| *Entropy levels at the input* | *21* |
| SYSTEM'S ADMINISTRATION | 22 |
| *Process of identifying the objectives of a system* | *22* |
| *Process of system management* | *22* |
| *Self-learning* | *23* |
| *Self-organization* | *23* |
| *Legalization of the system* | *23* |
| CONTROL SYSTEMS | 24 |
| *Indices of control systems* | *24* |
| *Feedback and control systems* | *25* |
| *Subsystems of Control* | *25* |

## 3. SYSTEM DYNAMICS   29

| | |
|---|---:|
| BASICS | 29 |
| *What is system dynamics?* | *29* |
| *Dynamic system* | *29* |
| *Synergistic or influence diagrams* | *29* |
| *Forrester diagrams* | *34* |
| *First-Order Dynamic Systems* | *37* |
| *Second-order dynamic systems* | *38* |

## 4. COMPUTER MODELING FROM SYSTEM DYNAMICS   41

| | |
|---|---:|
| COMPUTER MODEL STRUCTURE | 41 |
| *Structured programming* | *41* |
| *Object-oriented programming* | *41* |
| TOOLS | 42 |
| *Programming language* | *42* |
| *DYNAMIC_SYSTEMS class in C++* | *42* |
| PROBLEM OF THE CHICKEN POPULATION | 43 |
| *Mathematical model* | *44* |
| *Computer model based on C++* | *45* |
| *Computer model based on Java* | *46* |
| PROBLEM OF THE POPULATION OF ADULT RABBITS AND YOUNG RABBITS | 47 |
| *Statement* | *47* |
| *Objective* | *47* |

| | |
|---|---:|
| *Synergistic diagram* | *47* |
| *Forrester diagram* | *48* |
| *Mathematical model* | *48* |
| *Computer model based on C++* | *49* |
| *Computer model based on the class DYNAMIC_SYSTEMS* | *50* |
| *Java-based computer model* | *51* |
| PROBLEM OF POOL OF WATER TO A TEMPERATURE | 52 |
| *Statement* | *52* |
| *Objective* | *53* |
| *Synergistic diagram* | *53* |
| *Forrester diagram* | *53* |
| *Mathematical model* | *54* |
| *C++-based computer model* | *54* |
| *Computer model based on the class DYNAMIC_SYSTEMS* | *56* |
| *Java-based computer model* | *57* |

## 5. CONCURRENT COMPUTER MODELING FROM SYSTEM DYNAMICS   61

| | |
|---|---:|
| INTRODUCTION TO CONCURRENT PROGRAMMING | 61 |
| *Basic Definitions* | *61* |
| *Principles of Concurrency* | *62* |
| *Characteristics of concurrent processes* | *62* |
| *Concurrency problems* | *63* |
| CONCURRENCY IN JAVA | 64 |
| *Introduction* | *64* |
| *Thread Class Methods* | *64* |
| CONCURRENT COMPUTER MODEL STRUCTURE | 64 |
| *Structured Concurrent Programming* | *64* |
| *Object-oriented programming* | *64* |
| *Object construction* | *65* |
| PROBLEM OF SAVINGS ACCOUNT | 67 |
| *Statement* | *67* |
| *Purpose of the system* | *67* |
| *Synergistic diagram* | *68* |
| *Forrester diagram* | *68* |
| *Mathematical model* | *68* |
| *Computer model* | *68* |

## 6. COMPUTER MODELING CLIENT/SERVER FROM SYSTEM DYNAMICS — 75

INTRODUCTION TO THE CLIENT SERVER PROGRAMMING — 75
- *Client Server model* — *75*
- *Essential infrastructure components Client / Server* — *76*
- *Advantages* — *77*
- *Disadvantages* — *77*

JAVA CLIENT-SERVER PROGRAMMING — 77
- *Introduction to TCP/IP* — *77*
- *Communication via TCP* — *78*

CLIENT-SERVER MODEL STRUCTURE — 81

PROBLEM OF ONLINE SAVINGS ACCOUNT — 81
- *Statement* — *81*
- *Purpose of the system* — *82*
- *Synergistic diagram* — *82*
- *Forrester diagram* — *82*
- *Mathematical model* — *82*
- *Computer model* — *82*

## 7. SYSTEM DYNAMICS TO UML — 105

INTRODUCTION — 105

PHASE ANALYSIS IN UML — 106
- *Conceptual model of the system* — *106*
- *Use cases* — *109*
- *Sequence diagram of the analysis phase* — *111*

UML DESIGN PHASE — 114
- *Sequence diagram of the design phase* — *115*
- *Class diagram of the design phase* — *118*

**BIBLIOGRAPHY** — 127

# By way of prologue

Probably because of our helpful, warm, rising, uninterrupted and stimulating relationship we have built over the past 12 years, "referenced" by the exciting academic activities common to our work in this journey full of feverish, dramatic and diligent actions linked to the "production" of the top engineers in these technologies, modern and captivating, which unquestionably TIC's Clothing and dragging the world of today and tomorrow, the author of the present text, likely fueled by his proverbial generosity, I said the task to concoct some quick and brief way introitus dimensions to its intense, continuous and meticulous scholarly work wonderfully framed in the construction of relevant knowledge and its transmission to the generations that are to mark, to structure and guide them with a vision of transcendence towards society, in these appalling disciplines.

Going against the conventional schemes set out in this kind of exercise, when induced to future readers, whether they be students, teachers, researchers, professionals related to these disciplines guided by the latest technology framed in the so-called knowledge society, preceded the the Information Society that erupted in 2000 with the explosion of the Internet that chain came after the development of post-industrial middle of last century and the Industrial Society itself with identifying or if you want the discovery of electricity in the early nineteenth century who revolutionized factory life and woke up buried artisanal production schemes and goods, which are feeding and which serves the players and leaders, I will not refer to the text that the author makes the demanding close scrutiny of the academic community: The 152 pages, seven chapters covering the General Systems Theory-Foundations for General Systems Theory, Fundamentals of Systems, System Dynamics Computer Modelling, Computer Modelling Concurrent, building models Client-Server Computing conceptual and philosophical essence of the proposal linked inexorably to the aspect of "neural" in the corner over the formation of a systems engineer with a holistic view, we, as it has not be-a who will address the study of General Systems Theory and the training framework, and in the research, now in the discussions and controversies own fascinating scholars from these disciplines, when touched and become involved in dissecting the consideration of proposed text society of scholars of Informatics and perhaps in other disciplines concurrently.

I wish if, draw up a quick touch on the unusual and important features that make it possible to approach the personality, its own profile and hence to their work. the author of the text commented: Born and educated in his early years in a rural par excellence, the municipality of Turbaco settled in the outskirts of Cartagena de Indias, parents from this population and Maria La Baja, also rural, who distinguished themselves in their own work devoted to the support and development partner of six children, with jobs in the National Police as well as in the wholesale of beer the man, and the development of the home itself and the woman responsible for marking tip as the last child of this large family. Sent to the provincial capital, attended high school in the Salesian school-very-distinguished academic advantage over those years among the most remarkable for its promotion and evidenced a singular coincidence, which was the school where he would attend his secondary education and where he graduated as such, acted as a pioneer-to 1986-in available at that time a computer room in pairs exist at the time, slipping from those early years his curiosity, innate ability and interest in this discipline, just emerging in and strengthened our country in terms of calling for a close family friend, Don Jose Rodriguez or "Pepin" - who presciently sensed to be applied where the engineer in power.

The physical proximity and greater security in those years when the country was fighting a bloody battle against drug lords-1989 with the assassination of Luis Carlos Galan, 1990 and Beyond ", it was determined that the race was taken at Universidad del Norte disregarding the Industrial University of Santander (UIS) highly qualified and attractive to venture into this still new and perhaps unfamiliar discipline.

There, at the Universidad del Norte, the author forged his training of undergraduate and graduate student always standing on both journeys and Distinguished Fellow at the undergraduate as well as by the educational institution as the agreements with companies in the region stimulated talent as it was the foundation of such Mario Santodomingo conglomerate. In this, the Universidad del Norte, began studying for his growing and continued interest in systems theory and a paradox was not her guardian who induced him to travel on this career path because their teachings have never been to his liking, more if the subject itself began to interest you and excite since.

After working a short stay in Cartagena de Indias, and recent graduate of the first stage of their careers, linked from the beginning and continuously breaking our institution as a professor in the Systems Theory course, and soon after all disciplinary areas-career, this topic has made it since then and which has grown slowly and steadily, leading to the production of the text is now given to the academic community for tasting and rigorous scrutiny.

Now in its capacity as a researcher, his intellectual production has been devoted to develop and strengthen three specific topics, in which the Faculty of Engineering and the Foundation San Martín University's headquarters in Puerto Colombia, which through its umbrella of Dr. Jose Santiago Alvear has sponsored the publication of the book object of these dimensions with strong leadership and growing our strong academic and research activities, "mother and guide its tasks has been fortunate with enthusiasm and tenacity led to growing through its evolution: Theory Systems, Methodologies on Learning Objects, and Information Security.

Here's an incontrovertible paradigmatic example of this attractive development, dynamic and vigorous in this exciting production traffic on academic, intellectual and "engineering" of this wonderful specimen, who is part of our quarry prominent academic in the formation of the best and most transcendent engineers of these brand new generations, who will surely contribute to the transformation and better being of our fellows with a witty and creative technology solutions.

And, in a parody of the jargon so fashionable dining by these Kalends, I predicted "Bon Appetit" its lucky readers.


Jorge A. García Torres
Dean Faculty of Engineering
Fundación Universitaria San Martín, Puerto Colombia headquarters

Barranquilla, April 29, 2010.

# BASES ON THE GENERAL SYSTEMS THEORY

Chapter 1

## REDUCTIONIST APPROACH

### Specialization

We say that a ***specialist*** professional knowledge is highly deepened when studying a small area of knowledge. That is, a cardiologist who is a specialist health be trained in good shape to solve problems concerning the human heart, and a lawyer, a law specialist, will help tackle problems judicial nature.

Specialization has entered the area of knowledge and society with great force, replacing the "Wise Men" of antiquity. Comparing the elementary schools of our parents and grandchildren, are in the first, a teacher who taught all subjects (biology, languages, mathematics, aesthetics, physical education, etc.). But in the latter, subjects are given by various teachers. Similarly, when we consulted a "generalist", a disease that afflicts us, we often "refer" to a specialist in a particular area of health. To where it wants that we watch, we found the specialization, in the work, the schools, the universities, etc. Thus, for the development of any project, specialists of different areas from the knowledge "join" themselves to develop it.

The knowledge areas that represent the expertise are those that focus on a "part" of other areas of knowledge, for example: Each of the Health Sciences (Dermatology, Urology, Histology, etc.), And Engineering (Mechanical Systems, Civil, Electronics, etc.). With specialization, the term Master Integral completely disappears to make room at the end ***specialist.***

### Reductionist theory

***Reductionist Theory*** is a methodological approach based on specialization. That is, this theory studies the complex phenomena based on the analysis of its parts [1]. This theory focuses on move from general to particular, and when we have a toothache, we went to dentist (Specialist in Human Teeth) and not the Dermatologist (Specialist in Human Skin).

---

[1] Johansen, 1996

In fact, all undergraduate programs at universities are specializations of the total knowledge: the Masters are specialized studies of undergraduate programs; including doctorates so are the Masters and Postdoctoral are of doctorates.

It is known for the great contribution he has made to human knowledge reductionist theory, including the proper treatment of diseases, telecommunications, computer, etc., but it is also true that we do not enjoy the entire show to "specialize" ie, by reducing our object of study too.

Consequently, phenomenons exist like they are the **Computer Systems** that require to be analyzed as entireties, without losing of view the internal relationships; and they are not appropriately tried by the reductionist theory. In this type of phenomenons cannot be "known" neither to "predict" their behavior with the simple study of one on their behalves. For example, in the process of definition of the requirements of a software that must be built, it is impossible to determine them with the simple vision of a single user. It is necessary to keep in mind to all the users and clients, and also the relationships between them and their own necessities. When we analyze phenomenons with these characteristics we can fall in the imprudence and to generate missed knowledge and / or broken into fragments, originated the demand of additional resources to amend the error.

If we take as analyzes the behavior of a person in a given population, and we always results in that person complies with the traffic signals, would it be valid to say that everyone in the population also respect the traffic signals?

The great disadvantage of the reductionist theory is to generate **Specialized Ears** of specialists who have little communication with other disciplines, due to their knowledge so special. When their ears are more specialized, the lower your participation in a conversation between two or more professionals in different fields to study the same phenomenon. This is the case of a "conversation" between a lawyer and a physical star on worm holes.

## THE APPROACH OF GENERAL SYSTEMS THEORY

## Approaches to general systems theory

### System: Preliminary definition

For now, a **system** is defined as the set of parts that interact together to achieve a goal. Own this definition would be: football teams whose objective is to score more goals than your opponent; a refrigerator, parts of which relate to maintaining a temperature within the same; and the human digestive system which aims to transform appropriate energy the food humans eat.

*Methodology of General Systems Theory*
The **Methodology of the G.S.T.** is based on the analysis of phenomena as wholes consisting of parts interacting with each other (systems). Also aims to integrate the analysis of the phenomenon parties to reach a logical whole, where, are important relations between them. Therefore, we argue that the G.S.T. has a methodological basis contrary to the reductionist approach[1].

In G.S.T. the study objects are treated as systems and therefore seeks to overcome the disadvantages of the reductionist theory, forming the so-called **Generalized Ears**, and developing a framework containing a common language and to allow two or more specialists from different disciplines together to analyze a phenomenon. That is, these generalized ears will be able to "defend" in a communication for teamwork.

With this, the G.S.T. creates a new system, consisting of Generalized Ears (Parties) that communicate (interact) with each other, to analyze a phenomenon (Target). The situation is reflected in the case of a Working System for the construction of an information system, where the Software Engineer, Engineers from other disciplines, administrators, etc., must have the "protocols" right of communication for software development.

*Approaches of other authors*
**Von Bertalanffy**[2] defines G.S.T. as a logical-mathematical area whose mission is the formulation and derivation of principles that are applicable to systems in general.

For **West Churchman**[3], the G.S.T. is a way of thinking about systems and their components. In studying a phenomenon must first identify the objective pursued, and only after its structure.

## Frameworks for the study of G.S.T.

To implement the fundamental concepts of TGS in the analysis of the phenomena must choose one of the benchmarks described below:

*First frame of reference*
The **first frame of reference** is to construct a theoretical model to represent general phenomena that are in different disciplines. In fact, seeks in essence conceivable systems reduce to a manageable number. For example, in all areas of human knowledge there are populations of individuals; the idea is to generate a model that is applicable and valid in the different disciplines that have to do with those populations.

---

[1] To confront with Latorre,1996 and Johansen,1996
[2] Von Bertalanfy, 1978
[3] Churchman, 1973.

This framework provides an objective low ambition but with a high degree of confidence, trying to discover similarities in the theoretical constructions of different disciplines of knowledge, and developing theoretical methods applicable to at least two areas of study.

### *Second frame of reference*

The second frame of reference consists of a hierarchy of knowledge disciplines in relation to the organizational complexity of their components at a level of abstraction.

This second framework, presents a target of high ambition and low confidence, it seeks to develop a set of interacting theories (System of Systems) in particular areas of human knowledge, directing research to fill gaps. Table 1 describes this System of Systems.[1].

## Trends of practical application of G.S.T.

Among the trends of practical application of general systems theory are the following disciplines: cybernetics, information theory, game theory, decision theory, systems engineering.

### *Cybernetics*

**Cybernetics**[2] is the science that studies the transfer of information to the control and organization of systems. It uses the principles of feedback and homeostasis[3]. The objects of study of cybernetics are called **cybernetic systems**, which have parts that promote and manage the organization and control within the same, to maintain a balance of the system.

The typical example is the human central nervous system, for informing the brain to make a sudden movement of the right hand is burning, it acts as a cybernetic system, because with this action prevents the imbalance of the system

### *Information theory*

**Information Theory** is the science that is responsible for reviewing the handling that gives information as a contribution to the organization and implementation of the objectives of the systems. Looking for an Accounting Information System, which has worked well for several years, but at some point the government has enacted new legislation amending the methods of payment of taxes, this information must be handled properly to to keep "alive" to the System. Hence, all information affecting a system must be taken into account to generate new information and actions that impact the survival of the system.

---

[1] To confront with the description realised in Johansen, 1996
[2] Cybernetics. Developed by Norbert Weiner. Cybernetics. Cambridge Mass MIT Press. 1961
[3] Homeostasis. It is the property that presents the Systems to stay in balance.

### Table 1. Order of Hierarchy of the empirical Fields

| Level | Examples |
|---|---|
| **Static systems:** They correspond to conceptual or theoretical systems | The Conceptual Models<br>The laws of Newton<br>Trigonometry |
| **Simple Dynamic systems:** They correspond to nonorganic systems that transform some type of energy | Solar System<br>The Volcanos<br>The Sea currents |
| **Cybernetic Systems or Control:** They are Systems that help others to fulfill their objectives. | The Thermostat<br>The Human Nervous System |
| **The Dynamic systems first Order:** Systems with a first degree of organization. | The cells<br>The Virus<br>The Bacteria |
| **The Dynamic systems second Order:** | The Flora generally |
| **The Dynamic systems 3° Order:** | The Fauna generally |
| **The Dynamic systems 4° Order:** | The Man |
| **The Dynamic systems 5° Order:** | A Company<br>A family |
| **The Dynamic systems 6° Order:** | The absolute thing |

*Game theory*

**Game Theory**[1] is the science, using mathematical models, study skills or clashes between various systems capable of "reasoning" in which each participant seeks to minimize system losses and maximize profits. Among the cases studied game theory are: Fighting Sports, suppliers of a product on the market (as War of soda), the strategies of two men trying to conquer a lady and a police pursuit.

---

[1] Developed by Von Neuman and Morgenstein

### Decision theory

**Decision Theory** is the science of fighting between various systems, where some are able to "reason" and others unable to do so, in addition, each participant system capable of "reasoning" seek to make decisions that optimize the results (to minimize losses and maximize profits). Therefore, we conclude that decision theory is a special case of the Theory of Games, where players are rational. The example that stands out the theory of rational decision as no participant is nature. Among the phenomena studied for Decision Theory are: methods to mitigate forest fires, the management of supply and market demand, and prediction of weather and earthquakes.

### Computer Systems engineering

Carlos Trujillo[1] defines the Computer Systems Engineering as a discipline that aims to plan, design, test and build complex systems using the G.S.T. and engineering, as distinguished from the others Engineerings at its most integral character to examine the solution of problems.

Oscar Johansen[2] believes that Computer Systems Engineering concerns the planning, design, construction and scientific evaluation of human-machine systems.

For the author, Computer Systems Engineering is responsible for resolving problems, building automated information processing systems, under the approach of General Systems Theory using resources provided by engineering.

## FOCUS ART TO SOLVE PROBLEMS

In this section describe two approaches used in general systems theory in solving problems[3]. In the first instance, describes a formal procedure in which everything revolves around around the construction of models, and the second round of creativity. But first, let's define the concept of problem:

## What is a problem?

Is defined as a **problem** in the abstract difference is obtained by comparing the objectives with the results. Framing in the G.S.T., we can say that every system has goals to accomplish, if your product is different, conceptually to the objectives, it is said that a problem exists.

That is, for example, when a company does not have the right information on time, it produces can not make the right decisions or prevent mishaps, since what is desired (target) is to have all the information possible and have is uncertainty (results).

---

[1] Trujillo, Carlos. Análisis de sistemas. Mimeografiando. Universidad del Valle Colombia.
[2] Johansen B., Oscar. Introducción a la teoría general de sistemas. Limusa. México. P 32
[3] Ackoff, 1998

# First approach: modeling from reality

This **approach to solve problems**, describes a technique that involves the following steps: Identify the problem, decision to resolve the problem, Models of Reality, Use and work with the model and guidelines for action, decision, Commissioning, Operation and evaluation.

*Problem identification stage*

At this stage, which is seeking System objectives are not being met, making it clearly highlighting the magnitude and characteristics.

For example: In a grocery store a client requests a certain amount of merchandise which after having paid the shopkeeper realizes that there is no stock. The problem here is that there is no inventory control of merchandise.

*Decision to resolve the problema stage*

At this stage is performed feasibility analysis and decide whether "worth it" to solve the problem. To make the decision to solve the problem is necessary to conduct a feasibility study, which may cover several aspects such as:

- **Economical.** The question is if you have the resources necessary to fund the solution of the problem.
- **Technology.** It considers whether the technology exists to help solve the problem.
- **Operational.** It is important to know whether the proposed solution is applicable, used and accepted.
- **Motivation to solve the problem.** It is vital the actual arrangement to solve the problem.

In the event that one of these aspects is not feasible, you should not consider starting the process of solving the problem.

*Models of Reality stage*

The central idea of this stage is to model the behavior of the problem itself, guiding the knowledge of the situation and determine the overall objectives. Also, make the description of the system, identifying its supesystem, its subsystems, hierarchy and relationships.

*Use and work with the model and guidelines for action stage*

The model created in the previous stage is used for options of operation, to be able to define alternative solutions and evaluating them.

***Decision stage***

At this stage a group of people deal with the actions to follow. The decision may be to accept the proposals given by the study.

***Commissioning stage***

Is to plan and organize all the activities and tasks under the proposal accepted in the previous stage.

***Operation and evaluation stage***

This step ensures that the system works or operate regularly. In addition, it verifies compliance with the stated objectives through indicators.

## Second approach: creativity and constraints

In modern life, a professional in the area of General Systems Theory, must possess an essential feature that allows you to overcome obstacles and do not be average. This feature is creativity. Many authors argue that creativity is innate and therefore can not be taught or learned. The truth is that every person is born with some degree of creativity that must be developed with proper training from an early age.

Because oddly enough, the creativity of a person is mutilated by the type of education they receive from an early age, where, are taught to students to "think" in accordance with the guidelines of the school, family, country, etc thus suppressing the impulses born creative. By limiting creativity, ensures that institutions and models do not collapse. Thus, the wrongs of humanity are justified to maintain concepts that are the base of the institutions.

In his time, Galileo developed through research, mathematical modeling and observation, the theory that the earth revolved around the sun, it contradicted the arguments "accepted" at that time. Galileo used his creativity and solved a problem in way differently and correctly. Agree at that Galileo was right was to sow distrust of the believers who lead to the establishment of the disaster.

We thought then that if the children at an early age causes them to analyze and question the institutions, dogmas and paradigms, it is certain that revolutionary changes, innovative and useful they would more often when children they are older. It is also true that one way of doing things stops creativity.

For example, a math teacher puts on a review exercise that can be done in 5 different ways, but calls that are carried out by the method he knows. Truth be told, this teacher is teaching only knowledge that it dominates, moreover, does not allow students to develop other ways to solve the exercise, limiting possible learning first, and second, refusing to learn it from their students.

On the other hand, when we meet a group of friends and they say a riddle to solve, many, if we did not know before, we can not solve. This is the result that there is a self-imposed restriction, for example, has the following riddle: How would take a gold ring in a cup of coffee using just one hand, that the ring comes out clean.

In fact, the responses of the group of friends went from silly to ridiculous. All revolved around how to evaporate the water for coffee. The truth is that the solution to the problem was simply to get the ring with one hand full cup of coffee, because coffee is a solid and therefore can not get wet. The self-restriction placed friends restricted their creativity but this was so easy to apply.

We conclude that creativity is limited by self-imposed restrictions; therefore, to "get" Creativity should develop an ability to identify self-imposed restrictions and eliminate them. Clearly, to creatively solve problems is not sufficient to identify the self-imposed restrictions need a stronger boost.

# FOUNDATIONS OF SYSTEMS

## Chapter 2

## BASIC DEFINITIONS

### Definition of Energy

**Energy** is defined as material resources, financial, human and information that is transformed by a system to try to achieve their goals. For example, the furniture factory making wood (energy) and transforms it into chairs (energy produced).

The term energy was not taken arbitrarily to designate the inputs that are imported and exported products of super-Systems. The reason is that these energies meet **Universal Law of Conservation**, ie the amount of energy that belongs to the system is equal to the amount imported minus the amount of exported energy.

But there is an energy that does not comply with the Universal Law of Conservation: This is the **information**. That is, the information pertaining to a system is the difference between information entering and leaving, if this happens to teach my class General Systems Theory, the knowledge they impart to my students would necessarily have to forget, and in fact what happens is quite the opposite. To impart my knowledge, my students "store" this information, and I them, I can get more information, thus increasing my knowledge.

The great importance of the information in the General System Theory is this peculiar behavior which entitled the **Law of the increases**[1], which holds that the amount of information pertaining to the system is equal to the information that already exists, more information incoming, from there concluded that a system never deletes information. Therefore we can conclude that this is the reason you can not study the systems with the reductionist theory.

### Definition of system

In the previous chapter system was defined as the set of parts that interact to achieve a goal, now, nourish this definition:

---

[1] Ibid

***A System is a set of subsystems (smaller systems) that exchange energy to transform it (achieve a goal).***

Consider the Family system. Which among other parties is made up of parents and children, which in turn are also independent systems. In addition, any phenomenon is considered as a system when its constituent parts interact with each other and each are also systems.

For Javier Aracil[1], a system is a set of related parties functionally interdependent, which considered mainly interested in global behavior.

## Definition of Mega-system

Is defined as **Mega-System** or **Universal-System**, the system that contains all the systems in the universe. In simple words, all systems known to man, created and unknown, that are interacting with each other to form this great system of reference.

## Definition of Super-System

The **super-system**, a system of study corresponds to that system consists of all systems with which it is related that belong to Mega-system. For example, in the super system of a computer program would include the users, the computer, operating system, etc. The Super-System is the Union of all systems of the mega-system containing it. That is, every system which is part is included in the Super-System.

To identify the Super-System of a system takes into account all the systems with which it interacts. As each system is related to different systems of Mega-system for each system there will be a Super-System different, therefore each is unique Super-System. In addition, the Super-System can change over time, enough for the system to stop or start to relate to some other system to be modified.

Although the Super-System is unique and mutant, all have the same basic features of the systems (discussed below) that make it applicable General Systems Theory.

## Definition of SubSystem

**Subsystem** is defined as all those systems that make up the whole (or system) of study. The subsystems are classified according to the importance of the relationship with the aim of study, relevant and not relevant. The first, called **subsystem-Own**, are actively involved in achieving the objectives of the system, and the latter are treated simply as constituent parts. It is somewhat difficult

---

[1] Aracil, 1996

to determine if part of a system is a subsystem owned, so we suggest verifying the performance of any of the following rules[1]:

- **The Production Function.** That is to transform energy or provide a service. Presents an objective related to technical efficiency.
- **Support Functions.** Which would provide raw material for processing. For example, the departments of Public Relations and Marketing company
- **Maintenance Functions.** Its aim is to keep the parts of the system within it.
- **Functions of adaptation.** Its aim is to make the necessary changes so that the system can survive in the environment. For example, feasibility studies, reengineering, total quality processes, loss control, market research, etc.
- **Functions of Management.** That is to coordinate and plan activities and processes of the remaining subsystems also makes decisions

# ELEMENTS OF A SYSTEM

*Elements of a system* are all the relevant characteristics that help to better analyze a system under study. The most important elements of a system are[2]:

- Objectives
- Synergy
- Recursion
- The input currents (Inflows)
- The process of conversion.
- The output currents (outflows)
- The provision of feedback (Control Element)
- Frontiers
- Environment

## Objectives

*The objectives of a system* are the reasons why it exists. Without goals there is no system. All systems have objectives that are to transform energy, and only differ in "what" that transform energy.

In fact, we consider that the generic objective of a system is to transform energy in others. For example, an information system transforms the data (energy) in the information for decision-making (power transformer), while an iron transforms electric current (energy) into heat (energy Transformed).

---

[1] Johansen, 1996
[2] To confront with Ibid

Bearing in mind that any system generates a transformed energy or product, the objectives represent the ideal product that any system must generate. Systems may have general and specific objectives, wherein the specific binding of the general form.

## Synergy

**Synergy** is called the set of relationships or interactions between parts of a system. Similarly, Synergy is the energy exchange between multiple systems. She describes the way energy is transformed subsystems to meet the objectives. Synergy describes and determines the presence of relationships between constituent parts of a system.

The fundamental concept of synergy lies in differentiating the sum of its parts (subsystems) of the whole (system). For example, if we placed a container with a certain amount of water, carbon, iron and other substances that make up the human system, this mix is not out walking or even performing activities of mankind.

In fact, two or more systems may be comprised of the same parts and yet be different through synergy. A case of this situation is to compare a human being with a dog, obviously saving measures and ratios, the two systems have the same organic compounds and yet are completely different systems, the discrepancy is that the constituent parts are related (exchange energy) differently.

Another case that represents the importance of this synergy is: Two companies dedicated to building software, which have the same organizational structure and personnel, but we see that one has better results than the other and that's because the reference levels Synergy organizational, so a proper synergy causes better results. We conclude that the synergy, it represents the organization of systems.

Finally, by not being able to explain a system (which has synergy) from the analysis of one of its constituent elements, is inapplicable here reductionist theory, there lies the usefulness of the GST, because it provides the method that helps understanding the system studied.

## Recursion

**Recursion** is called to the property that have the systems to be composed of elements (subsystems) that are themselves, behave and are studied as systems. Recursions provides subsystems for the characteristic of being separate elements, but in turn inherit the properties and principles applicable to the Systems.

Finally, recursion in systems expressing degrees of complexity and hierarchy. Thus, the human being is, among other parties, the human central nervous system, which in turn appears as the Sub-System Neurons are also systems.

*Consequently, we can again define the systems as a set of parts that have the features of Synergy and Recursion.*

## The input currents - Inflows

**The input currents**[1] are all energies that are imported from the Super-System. The input currents are the inputs or raw materials that the system needs to fulfill its objectives.

The energies that shape the input currents are the products of systems in the Super-system which relates the system being studied. On the other hand, the systems receive, through the input currents, the energy necessary and proper super-system, essential to its operation. ***Figure 1*** describes the input currents of a system.

The extreme dependence of a system to input currents generates major constraints and in some cases, when there is energy shortage is threatening their livelihood. This is why there are some systems that persistently fight for greater access and control over their energy sources. An example would the plants, which, when deprived of sunlight (placed under the shade of a building) can extend its branches until the leaves can access it, and they do because without it could not perform their basic tasks.

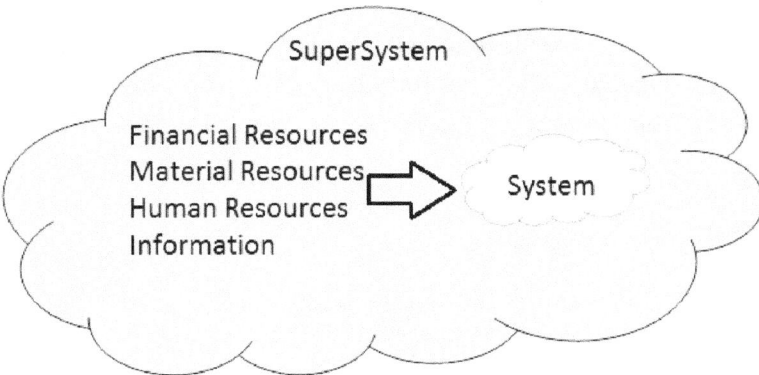

**Figure 1. The input currents**

---

[1] Confront with Ibid, Churchman, 1973 and Latorre, 1996

## The conversion process

The energy supplied from the super-system through the input currents are transformed so that the system can achieve its objectives. **Figure 2** describes the conversion process. Every subsystem of a system transforms the energy that is provided; this will be called **partial conversion** of the energy. In the end, these conversions will be transformed by partial special subsystems to refine and complete the conversion of imported energy.

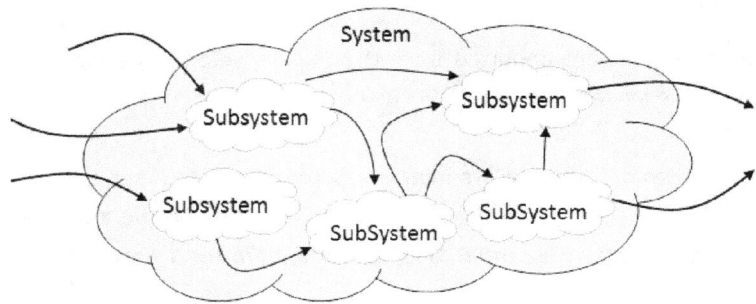

**Figure 2. The Conversion process**

## The Output currents - Outflows

The **output currents** corresponding to the products or processed energy, which the system under study exported to Super-System. The output currents are composed of a series of transformed energy, which are classified as *positive* because they are useful to the Super-System, or *negative* because they are not useful. **Figure 3** describes the output currents. Output current "smoke" is positive, but for those who do not smoke constitutes a fact or negative stimulus. With the above, we can say that the output currents of a system are evaluated by others belonging to the super-systems, in many cases from the perspective of their particular interests at the expense of their own legalization.

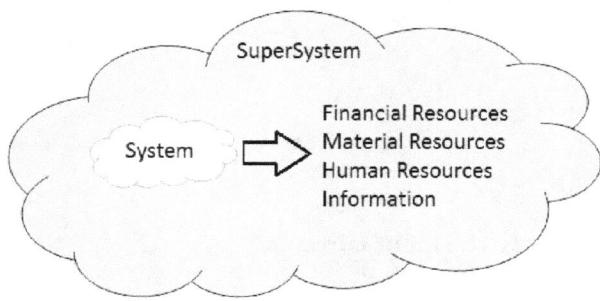

**Figure 3. The output currents**

## Feedback communication - The provision of feedback

**Feedback Communication** is the information that enters the system that allows us to see if the system is meeting its objectives. This information is obtained using a procedure that involves comparing the output current with standards that quantify the objectives of the system, further indicates the difference found corrective actions to be performed. **Figure 4** describes the Feedback Communication.

It is noted that the Feedback Communication is *information*, based on the analysis of output currents, which is introduced into the system to make the necessary adjustments to meet objectives. There are two types of feedback: Positive, when adjustments reinforce the momentum, and negative, that attenuated the initial effort. Positive feedback is used when the objectives of the system tend to infinity (+ or -). In contrast, the negative feedback is used when the objectives of the system are accurate.

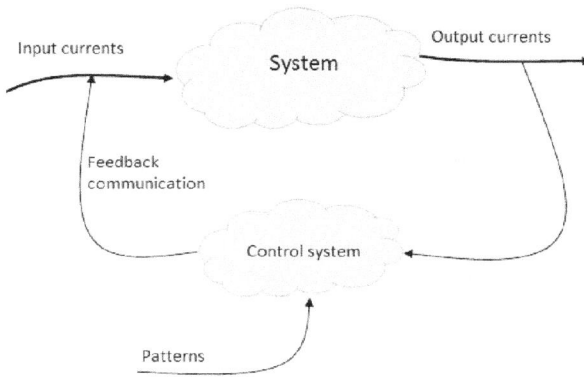

**Figure 4. FeedBack communication**

## Frontiers of the system

The **Frontiers of the system** defines what the systems of the Super-System belong to and what not. The Frontiers also define the structure of the system. There are 2 types Frontiers: Physics Frontier and Functional Frontier.

The **Physics Frontier** is one that defines a geographical area or space in which the system interacts. For example, the limits of a city and human skin.

The **Functional Frontier** expresses the limits in relation to the activities. For example, a Road Transport Company issued tickets and turns in their vehicles, but not men's clothing design.

## Environment System

The **environment of a system** containing all the parts and systems of the Super-System outside the studio system. Generally, the environment affects the system and the changes that occur in it, determine the behavior of the system significantly. The definition and identification of its environment is linked with the objective of the system, and point of reference for people who study it. **Figure 5** illustrates graphically the concept of a environment system.

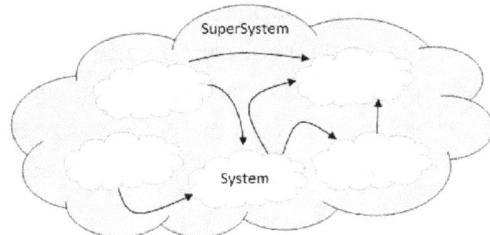

**Figure 5. Environment system**

*Active environment*

The **Active environment** of a system what are all the systems belonging to the super-system, that provide energy. That is, are all systems that relate to the system through its input current.

*Passive environment*

The **Passive Environment** of a system are all systems belonging to the super-system which they import energy flows out of the system. Figure 6 describes the two types of environment that has System

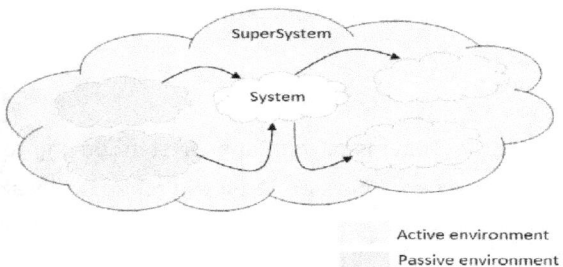

**Figure 6. Types of Environment**

## LEVELS OF ORGANIZATION OF SYSTEMS

Based on the concept of Synergy's own systems, we present the idea of organization structure Subsystem - System - Super-system. In fact, when you move from one subsystem to a system as an object of study has shifted to a higher level of organization and in turn, going from one system to a super-system, has moved to a much higher level of organization in relation to the subsystem. When it delves into an object of study, from a subsystem to a system and then to a super-system, the complexity of the object of study is more, as well as understanding their behavior[1].

Moving in a manner contrary to (reductionist approach)

**Super-system -> System -> Subsystem**

The whole is less information.

Figure 7 describes the relationship understanding, complexity, organization of the whole advance in the study vs. a super-system to a system and then to a subsystem:

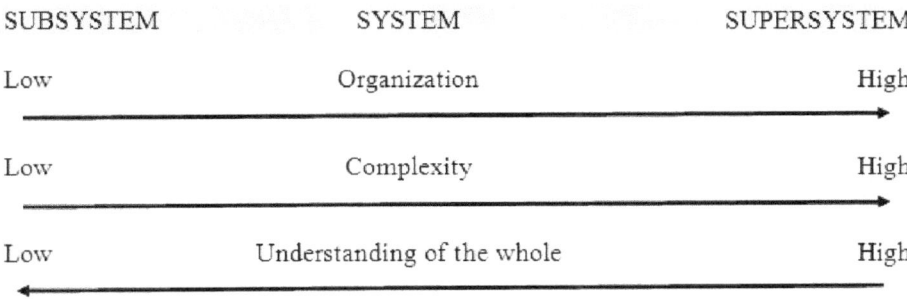

Figure 7. Organization Subsystem, System and Super-System

# ENTROPY IN THE SYSTEMS

## Entropy

**Entropy** is defined as the energy that go into a system will cause a continuous organizational change, reflected in passing from a more organized with a less organized, or what is the same, go slowly, to a less likely ( organization) to its most likely in nature (chaos). Clearly, if we for a long time a solitary house when we visit again, we would find it "falling", or at least leave in the open a number of bricks, over time we see that they are apart. This is the manifestation of the entropy leads to its organizational systems more likely: The complete disorganization.

All energy that matters the system, to make you into its constituent parts or in their relations, chaos and misinformation is considered Entropy. For example, a person who daily breakfast coffee with rum,

---

[1] Confrontar con lo expuesto en Churchman, 1973

there will come a time when this disease of the liver and central nervous system, as the rum and coffee are entropy and these are generators of clutter and disorganization. Entropy gives rise to diseases of the system and brings them to death. It is no secret that the best way to destroy a system is disorganized.

The Entropy has as function the one to destroy to the System and for that reason is our greater preoccupation. Thus, the system Human Being, presents throughout its life a physical deterioration, and organizational decline that at some time will take to the death.

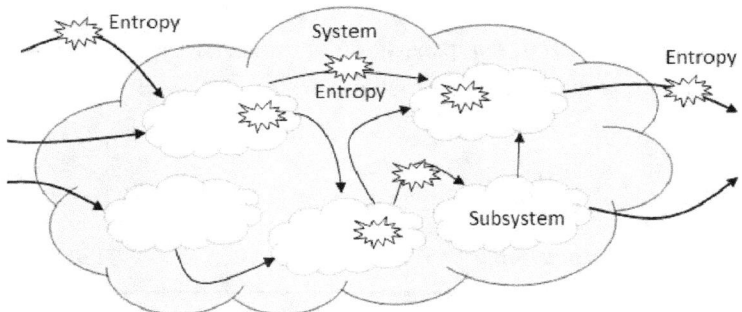

**Figure 8. The Entropy in the Systems**

The diseases that we presented the human beings are product of the Entropy that is accumulated in us. The Entropy produces a Cancer that mine little by little to the Systems, and most serious it is than this "disease" or tendency to the distortion presents all the Systems.

The chaotic effects and of disorganization that produces the Entropy are cumulative, beginning in a level of 0% when being born and the 100% when dying. The death of a System catalogues like the **Maximum Level of Entropy** and the birth like its minimum level.

It is necessary to consider that the Entropy always tries, no matter how hard we want to avoid it, to disorganize, to disinform and to create chaos in the Systems. It is for that reason that always is due to consider when we analyzed to a System.

## Negative Entropy

The **Negative Entropy** is the energy that when entering a system foments the improvement of the organization of its constituent parts, attacking the disinformation and the chaos. The main function of the Negative Entropy is the one to maintain low levels of Entropy in the systems, in this way "extends" the existence to him. Although it is impossible to completely eliminate the effects of the Entropy, the idea to maintain levels low of Entropy assures that the system operates to a 99%.

As well as the Entropy is the energy in charge to destroy the systems, the Negative Entropy is the energy that makes the birth possible of the systems. When dying a system, that is to say, their constituent subsystems no longer interact in search of an objective, in a time not infinitely, is created what **AntiSystem** is denominated.

This AntiSystem is conformed by Sub AntiSystems until certain hierarchic level, of there downwards, the Sub AntiSystems are formed by subsystems able to interact with other subsystems pertaining to Super-System being in freedom to form new Systems

On the other hand, we know that at some level of Entropy and its effects it will be always in the Systems. A system resists that it with the Negative Entropy, will be to him helpful for achieving its objectives. The energies that have organizing functions, of order and to inform are Negative Entropy, that prevents "the premature" death of any System. The function To inform inherent in the Negative Entropy makes think us that the Information is a "flowing source" of this type of Energy, but is necessary to consider that "to inform too much" can generate Entropy.

With base in the exposed thing previously, if one analyzes the characteristics of the Information, one concludes that it is an energy that looks for in its essence the order and the organization. For that reason the mathematical equality between the Negative Entropy and the Information is deduced.

## Entropy levels at the input

As much the Entropy as the Negative Entropy is introduced to the Systems through the constituent energies of the Input Currents, and it as well sends them to the system to its Super-System through the Output Currents.

In the Input Currents, there is a percentage of Entropy and another one of Negative Entropy, both adds the 100%. The minimum percentage of entropy that enters a System, never is zero, since all energy by natural law takes a desorganizador element, for that reason this minimum level is very near zero but never he is zero.

On the other hand, the maximum percentage of Entropy that enters a System naturally is the 100%. In order to describe the interval of levels a percentage scale will be used by comfort positive whole, from which tenth that the interval that represents the Amount of Entropy that enters a system is [1%-100%]. Starting off of the previous interval we concluded that the interval that represents the amount of Negative Entropy that enters a system is [0%-99%]

## SYSTEM'S ADMINISTRATION

### Process of identifying the objectives of a system

Previously it was defined what they were the Objectives of the System, now, analyzes a methodology for its correct identification. It is not easy to determine what is the objective of a System, since a standard methodology does not exist to identify them. But, the following one is suggested:

It is selected, first of all, a range of possible objectives that would fulfill the system to which we analyzed, remembering the energies of the exhaust streams. This range of objectives we will call **Objectives Candidates**.

Secondly, they are taken one by one, the Objectives Candidates, and it is analyzed if the system sacrifices the others to fulfill this objective, in positive case, this candidate is an objective of the system.[1] For example, a Countable Information system were defined the following Objectives Candidates:

- Connection to Internet
- To lend support of access to the disc
- To maintain the accounting to the day.
- To send reports to the Suppliers.

It is clear that the system would sacrifice all the other Objectives Candidates to "maintain the Accounting to the day"

The Objective of a System represents sumatoria of the objectives of the subsystems conform that it. In fact, the methodology explained here is valid for all system including from the MegaSistema, happening through Super-System and the subsystems.

### Process of system management

The **process of system management** is in charge to level Macro to verify the fulfillment of the objectives of the System; And at Micro level to verify and to make pursuit of the fulfillment of the objectives of each one of the subsystems of the System, with the purpose of to apply corrective the necessary ones when and where it is necessary. These functions are reserved to special subsystems that we will denominate **subsystems of Administration.**

---

[1] Latorre, 1996

The subsystems of Administration are the ones in charge to define the objectives of the other subsystems, as well as to provide resources, to organize and to control the behaviors of the system. An example of typical Subsystem of Administration is the Subsystem Brain in the System Human Being. Other functions of the subsystems of Administration are: the generation of plans, use of the resources, control of the profit of individual missions and total, and the legalization of the system.

## Self-learning

A System presents **Self-learning** when the subsystems of Administration are able to generate changes in the form as the tasks are realised with the purpose of to adapt their surroundings better, based on the happened experience.

A Software of artificial intelligence generates changes of "conduct" at the most is its use. For example, an Intelligent Software of Security often is proven with real delinquents in order that self-learns from real confrontations with human opponents.

## Self-organization

A System presents **self-Organization** when the subsystems of Administration are able to modify the structure of the organization in progressive form, with the purpose of to obtain its objectives. For example, is said that the System Factory of Shoes self-Organizes when it implements in his operation quality controls.

## Legalization of the system

The **Legalization of a System** is the "Visa" that allows him to import and to export energy to Super-System. All System owns a **Level of Legalization** which influences in the amount and type of energy can import and export to Super-System.

The Legalization of the System is its life in Super-System. The low levels of Legalization indicate that the system does not own the capacity or is not allowed him to concern the suitable energies to achieve their objectives, which represents their progressive degeneration.

A typical case of the legalization, is a Software "x" that is not friendly for the user. The engineer of Software resists to modify it, arguing reasons that he creates advisable. Consequently the people who would have to use it not do it, of there, the data that do not introduce themselves, they are not process and in the long run the lack of use finishes with the "life" Software "x".

In order to increase the Level of Legalization, a System must use its administration to modify its structure, that is to say, its administration must foment, direct and to verify the self-organization, in addition, must create and to apply its normalization of processes to arrive at a self-Control, additionally must generate the sufficient freedom in Super-Sistema like owning autonomy.

The Level of Legalization of a System considers like the degree of relation and interaction with its Super-System, which takes to us to determine that the Level of Legalization is not more than the degree Synergy of the System in relation to its Super-System.

## CONTROL SYSTEMS

All System must watch the fulfillment of its objectives, for these is important to develop the capacity of adaptation in its Super-Sistema. In order to adapt, a system must audit its "conduct" in relation to the own exigencies of the Systems that interact with him.[1]

What here we called conduct of the system is not more than to produce what Super-Sistema needs that it produces. One is due to consider that all system belongs to a greater system, than as well, needs that all systems suitably turn the provided energy.

Like example, an educative institution is had in which it is tried to foment the value of the ethics and the moral; all the educational ones that belongs to her must educate with the example, behaving with adapted ethical and moral when distributing their classes.

In the process of Control, the systems must be reinformar comparing their objective with the produced thing, and realise the adjustments necessary with the purpose of to reduce to the maximum the difference to reasonable terms.

### Indices of control systems

In the control of a system it is necessary to have parameters that they indicate at a certain moment if the system is fulfilling its mission, for that reason describe three (3) indicatives, that we denominated the EEE: **Effectiveness**, **Effectiveness** and **Efficiency**.

The **Effectiveness** of a System measures the profit of its specific Objectives. That is to say, the Effectiveness measures the difference between the product of the system with its specific objectives, between major is this less effective difference is the system. If we analyzed the System Warehouse of Shoes, presents a specific objective to sell per month a volume of 40% of the inventory and it only sells 5%, we found that this System is not Effective. But, if on the contrary the volume of sale is of 37%, necessarily we concluded that the system is Effective.

---

[1] To confront the exposed thing with Johansen, 1996

The **Efficacy** of a System measures the profit of its General missions. That is to say, the Effectiveness measures the difference between the product of the system with its general missions, between major is this less effective difference is the system. If we analyzed again, the System Warehouse of Shoes, presents a general mission to increase the sales in a 50% and it only obtains 1%, we found that this System is not effectual. If the increase is closely together or is superior to 50% definitively he is efficacious.

The **Efficiency** of a System measures the profit of its Objectives considering that resources and that Costs were used to obtain it. The idea is to achieve the objectives on the basis of the minimum costs or in their defect in "Reasonable Costs". When we analyzed system that obtains his objectives using great amount of his resources, this it is very struck and vulnerable, which influences that it does not continue operating suitably, we concluded immediately that it is not an efficient system.

## Feedback and control systems

With the Negative Feedback, the Systems tend to remain in balance. This characteristic is propitious to carry out an suitable control of the Systems. With the Positive Feedback the control is impossible; since the parameters change continuously, in addition, always lies down to eliminate the effects of all planning.

Thus, a student who in his first note obtains a qualification of 4,0 on 5,0, like this one is major that the required minimum note (3.0), then would be placed as a note puts obtaining, for the next examination, of 2,0 how minimum. Nevertheless, the student does not study much and in his second examination she obtains a qualification of 1.0. Then, now its goal is to obtain a qualification of 4,0 to gain the matter. It is observed in this case, that the objectives always changes as it realises a new examination, takes it to a disinformation, besides a total uncontrol. With the Negative Feedback the same student would determine an objective to obtain how minimum a qualification of 4,0 in each examination, and if obtains more or less, its study always would have a same level of learning. We find in this case a smaller variation of the interest level and study of the student.

## Subsystems of Control

The **Subsystems of Control**[1], are the parts of the system that are in charge to control to the system. The parts that constitute a Subsystem of Control are:

- Objective to control
- Subsystems of Sensitivity
- Subsystems Motors

---

[1] Ibid

- Resources of Energy
- Channel of Feedback

***Objective to control***

The ***Objective to control***, as to his it indicates it name, is one of the objectives of the system that needs control. It applies the studied thing, referring to the objectives identification.

***Subsystems of Sensitivity***

The ***Subsystems of Sensitivity*** are the ones in charge, first of all, to measure the changes provoked in the product of the system, and in second term, to realise the comparison with the patterns.

***Motors Subsystems***

The ***Motors Subsystems*** are the subsystems in charge to plan, to manage and to process the remedial actions.

***Resources of Energy***

The ***Resources of Energy*** are all those energies that need to be imported by the Motors subsystems to realise the pertinent corrections.

***Channel of Feedback***

The Channel of Feedback is the communication process between the subsystems of Sensitivity and the Motors subsystems, which transport the corrective information.

***Example:*** We consider the System conformed by Man - Radio and analyze a possible Subsystem of Control.

**Solution**

- ***Objective to control***. Quality of the sound that produces the radio. Which can be product of a bad tuning of a transmitter, statics (there is no a transmitter in the "dial"), damage in the radio.
- ***Subsystems of Sensitivity***. It is the auditory System of the human being.
- ***Motors subsystems***. In all the cases the average motors are the Muscular System of the human being, and in the case of the repairs by damages, the own tools radio technician.
- ***Resources of Energy***. They are the natural sources of locomotion of the human being, electrical energy, batteries, etc.
- ***Channel Feedback***. The communication channel is the air, using concretely its characteristic of propagation of the sound.

***Example:*** Now, the System is considered Thermostat and we studied a possible Subsystem of Control.

## Solution

- ***Objective to control***. The temperature of a room.
- ***Subsystems of Sensitivity***. Highly sensitive thermometers.
- ***Motors subsystems***. The Switch System, the System of Cooling, the system of extinguished automatic.
- ***Resources of Energy***. Electrical energy.
- ***Channel of Feedback***. The molecules of the room. The air of the room.

# *SYSTEM DYNAMICS*

Chapter 3

## BASICS

### What is system dynamics?

The **system Dynamics** is in charge of the study of the behavior and evolution of the systems through Models and Simulations. The system Dynamics creates an abstract or conceptual Model from which the dynamic behaviour of the real system is analyzed, to this model, Dynamic System is denominated to him.

### Dynamic system

A **Dynamic System** is the conceptual System or model in which they are formalized the parts and relations of a System to which it is tried to him to study his dynamic behaviour. That is to say, the Dynamic Systems are Systems of Modelaje whose main function is to describe its dynamic behaviour through time, on the base of the existing relations between its constituent subsystems and with the Super-System.

In this way, in the Dynamic Systems the subsystems that conform the system in study, like the systems of their Super-System consider as much to which it is related. First **Endogenous** subsystems are called and the second **Exogenous** subsystems ones.[1]

The Dynamic Systems belong to the hierarchic category of the Static Systems, to the being conceptual systems, whose objective is the description and prediction of the processes of import, transformation and export of the energy that enters the study system.

### Synergistic or influence diagrams

The **Synergistic Diagrams** are in which the names of the subsystems of the System in study appear united with an arrow with the purpose of to allow to know their structure in a Dynamic System.

In this structure the relations of the subsystems are specified. It is necessary to write down that the systems of Super-System that present relation with the subsystems, also are described in the

---

[1] Aracil, 1986

Synergistic Diagrams. The Synergistic Diagrams describe, in addition, the type of Synergy that exist between the subsystems.[1] The main function of the Synergistic Diagrams consists of describing existing energy exchange to them of the subsystems, with they themselves, and the other systems of Super-System, considering the objectives of each of them.

For example, we say that a system in study owns two subsystems H and C. If an energy of the exhaust streams of H is an energy of the input Currents of C, then, we can say that H is able to influence to C, we indicated and it in a Synergistic Diagram by means of shoots with an arrow leaving H towards C. (See Figure 9)

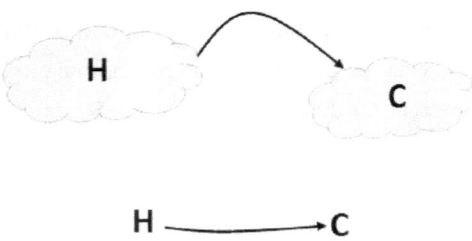

**Figure 9. Example of synergistic diagram**

The relations or synergies between the subsystems of a system in study have a positive or negative **sense** according to the following premise: The **positive relations** make reference to variations of profits of objectives between systems in a same sense, that is to say, that a positive relation of subsystem H exists towards subsystem C, when when increasing the profit of the objective of H, the profit of the objective of C also will increase, or what is the same, the more it achieves his objectives subsystem H, the more will help his product to that subsystem C obtains theirs; if on the contrary, if it diminishes the profits of objectives of H also in C will fall.

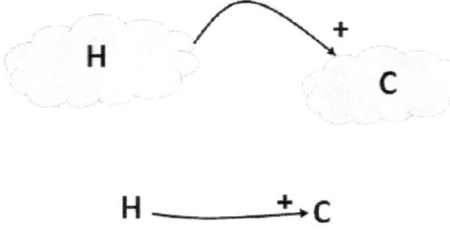

**Figure 10. Positive Relation**

---

[1] To confront with Ibid; Aracil and Gordillo, 1997

The **Negative** relations make reference to variations of the profit of objectives between systems in an opposite sense, that is to say, that a Negative relation of subsystem H towards the subsystem C, if increase the profit of the objective of H, the profit of the objectives of C falls, but, if the profits of the objectives in H falls, this causes that they increase the profits of the objectives in C.

Finally, the relations set out in the synergistic diagrams are of two types, a first relation of Cause - Effect, and the other of correlation, for example, the statistical relations

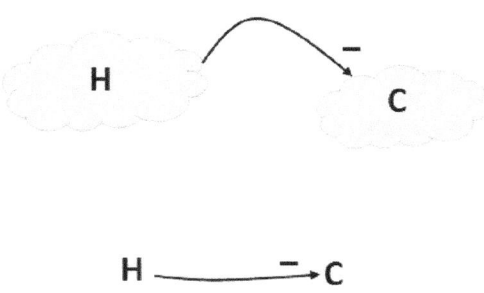

**Figure 11. Negative Relation**

### Types of Structures of the synergistic Diagrams

Exist two types of structures, the **simple** and the **complex** one: The **Simple structure synergistic Diagrams** are those in which they do not appear relations that represent closed ways.

For example, they are had related three subsystems A, B and C; where relations exist thus: a relation of the subsystem "A" with "B" and another one towards "C", but, not of "B" to "A", nor of "C" to "A", as is it shows in Figure 12.

**Figure 12. Simple structure synergistic Diagram**

The **Complex structure synergistic Diagram** are characterized to present closed chains of relations. (To see Figure 13) To these closed chains of relations one denominates **cycles to them of feedback**.

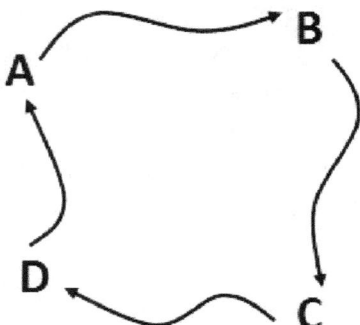

**Figure 13. Complex structure synergistic Diagram**

### Cycles of positive feedback

The **Cycles of Positive Feedback**[1] are those chains of relations in which an effect or variation of the profit of the objective propagates it in all the constituent subsystems reinforcing this initial effect. It is To observe that all Cycle of Positive Feedback it presents an even number[2] of relations Negative, because when realising a complete route of the cycle, the effects of a negative relation is neutralized by negative relation other, as it is described in Figure 14.

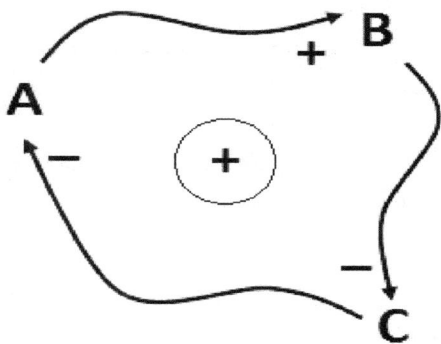

**Figure 14. Cycles of positive feedback**

Of Figure 14 we make the following analysis:

If it **increases** the profit of the Objective of **A**, then, it increases the profit of the objective of **B** (positive relation). If it increased in **B**, then it decreases in **C** (Negative relation). If it decreased in **C**, then in **A** it **increases** (negative relation).

---

[1] Aracil, 1978 , Aracil and Gordillo, 1997
[2] The zero (0) like even number are due to take.

If the profit of the Objective of **A decreases**, then, the profit of the objective of **B** decreases (positive relation). If it decreased in **B**, then it increases in **C** (Negative relation). If it increased in **C**, then in **A** it **decreases** (negative relation).

### *Cycles of negative feedback*

The Cycles of Negative Feedback are those chains of relations in which propagate an effect or variation of the profits of the objectives in all the constituent subsystems, with opposite sense in the same element. A Negative Cycle of Feedback presents an odd number of relations Negative. (See Figure 15)

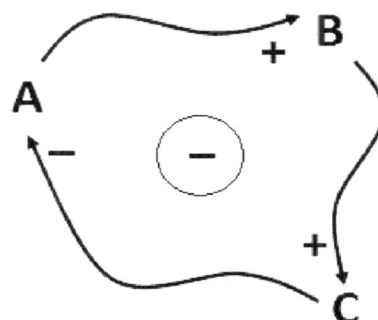

**Figure 15. Cycles of negative feedback**

From Figure 15 the following thing concludes:

If it **increases** the profit of the Objective of **A**, then, it increases the profit of the objective of **B** (positive relation). If it increased in **B**, then it increases **C** (Positive relation). If it increased in **C**, then in **A** it **decreases** (negative relation).

If the profit of the Objective of **A decreases**, then, the profit of the objective of **B** decreases (positive relation). If it decreased in **B**, then **C** decreases (Positive relation). If it decreased in **C**, then **A** it **increases** (negative relation).

### *Problem of the population of chickens*

An Agriculturist dedicates itself to the young of chickens in his property. Purchase 240 chickens, hens and roosters, apt to reproduce. By the experience of more than 20 years than it takes in the business knows that every month they are born between 2 and 50 chickens, in addition, die between 1 and 5 chickens. Every month the agriculturist sells between the 0 to 10% of the total population. Realise the corresponding synergistic diagram of the system. The synergistic diagram is described in Figure 16.

34  General Systems Theory: A focus on computer science engineering

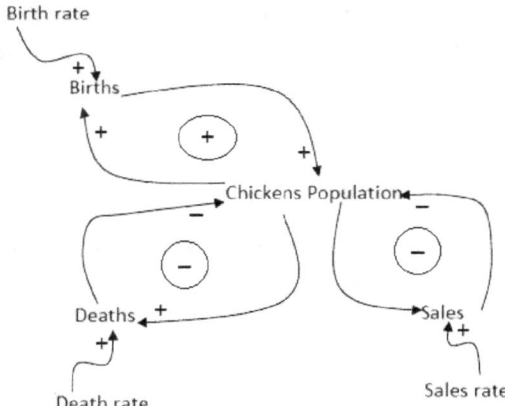

**Figure 16. Synergistic diagram of case of the chickens.**

## Forrester diagrams

With the purpose of to contribute to the Dynamic System an exact description of the operative nature of the subsystems and the relations that conform to the system in study, is represents to the elements that described in the synergistic diagrams, according to the functions which they realise, in Level Subsystems; Flow Subsystems; and Auxiliary systems. This graphical interpretation is to which we called **Forrester Diagrams**[1].

### *Channels of Material Energy and Information*
The relations between the subsystems of a system in study are realised through the transformed energy exchange. It is energy can be represented in material elements, like the wood and the steel, or in information like for example the one that is stored in a data base. In the Forrester Diagrams these energies represent by an arrow directed indicating the sense of the relation, that is to say, a channel of material energy is represents by an arrow continuous, while that the information channels is represents by an arrow interrupted.

### *Level Subsystems*
The **Level Subsystems** represent the subsystems in which they are accumulated energy, which can be physical or abstract. In order to identify to a Subsystem of Level is based on its characteristic to change or to store when interacting with other subsystems or parts of the system in study. In the Forrester Diagrams the levels is represent by rectangles.

Two Level Subsystems cannot be related through the material energy that they accumulate, since if a Level 1 stores fish and a Level 2 stores money, becomes difficult that it among them interchanges

---
[1] Ibid

these two different energies. But two Levels can be related through the referring information to which they store, such as: the stored amount, the variations happened in a time interval, etc. (See Figure 17)

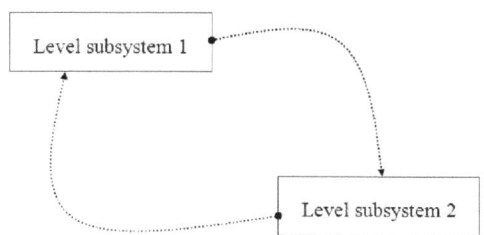

**Figure 17. Level Subsystems**

### *Flow Subsystems*

The **Flow Subsystems** represent the subsystems that determine variations in the Level Subsystems. The Flows are like valves that regulate the content of the Levels, is for that reason that the Flow Subsystems s are associates with decision elements with the purpose of to determine the type of variation which they will undergo the levels.

The Flows are related to the Levels by material energy or Information. In fact, a Level subsystem is related at least to a subsystem of Flow, but, two Flow Subsystems cannot be related directly by material energy. In Figure 18, if we observed the diagram of the right, it shows the incorrect form to relate two Flow Subsystems. However the diagram of the left shows the correct form.

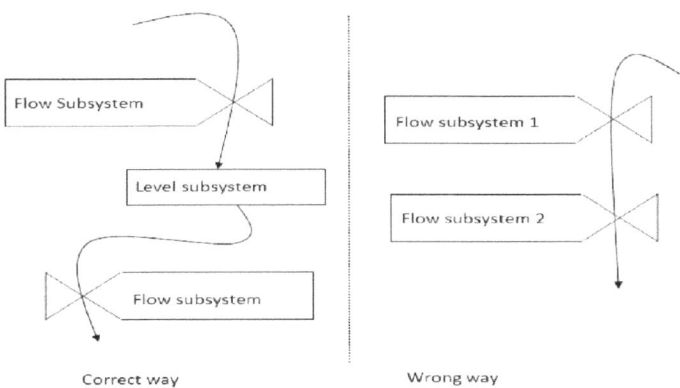

**Figure 18. Flow Subsystems**

### *Auxiliary systems*

The **Auxiliary Systems** represent the systems of Super-System that are related to the subsystems of the System to which its dynamic behaviour studies to him. In most of the cases the Auxiliary Systems

# 36 General Systems Theory: A focus on computer science engineering

represent the external systems or information channels between the Level Subsystems and subsystem of Flow. It is used the ovals for his graphical representation, see Figure 19

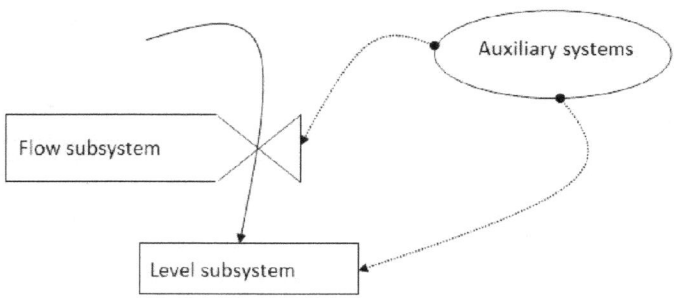

**Figure 19. Auxiliary systems**

***Other symbols in the Forrester Diagrams***

With a **Cloud** is represents a **source** or **well**, that is to say the relation of the system with Super-System. According is needed to a cloud-well will be the output currents and a cloud-source, will be the input currents. The **constant values** represent by **small circles cross by a line**, while the **random values** with a **small circle**. In Figure 20, the different symbols from the Forrester Diagrams and their meaning are detachhed[1].

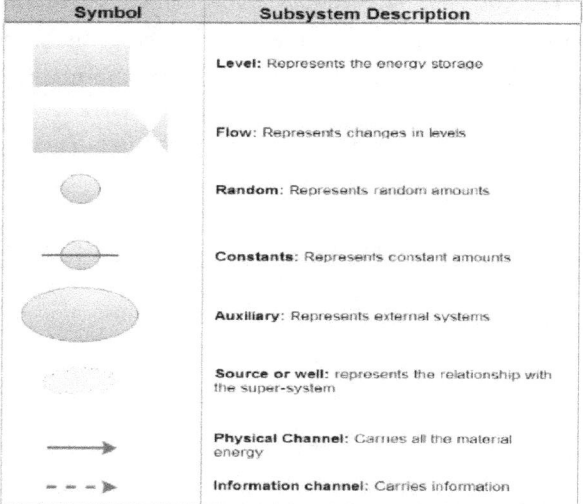

**Figure 20. Summary of Symbols of the Forrester Diagrams**

---

[1] Ibid

*Logico-mathematical model of the Forrester Diagrams*

The Logico-mathematical Model of the Forrester Diagrams is synthesized in the following one formulates:

$$\text{Present Level} = \text{Previous Level} + \Sigma \text{ IF} - \Sigma \text{ OF}$$

Where,

IF: Associated input flows (inflow) to Level

OF: Associated output Flows (outflow) to the Level

Typically the formula is read: The present state of a level is equal to the previous state plus sumatoria of the influences of the input streams minus those of output.

## First-Order Dynamic Systems

The **First-Order Dynamic Systems are** characterized to present a single Subsystem of Level in their structure. In a dynamic system of first order appear cycles of feedback of two types: positive and negative.

*First-Order Dynamic Systems with negative feedback*

Dynamical systems are those that have a behavior over time, allowing you to maintain balance in your subsystem level. As an example we have the thermostat system, whose main function is to maintain a constant temperature in a room. In Figure 21 describes this behavior over time.

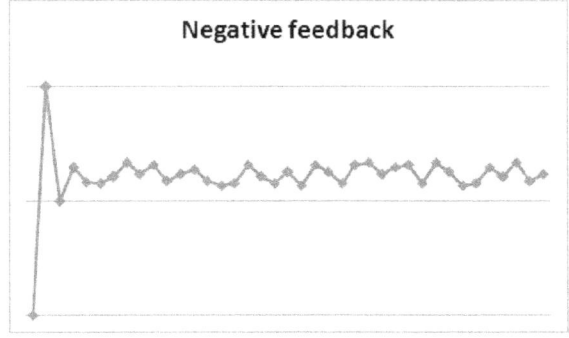

**Figure 21. Negative feedback**

*First-Order Dynamic Systems with positive feedback*

Are those with behavior that creates a growth or a rapid decline of the subsystem level. With them the accumulated levels tend to go to extremes, that is, to zero or to stop. In Figures 22 and 23 describe these situations.

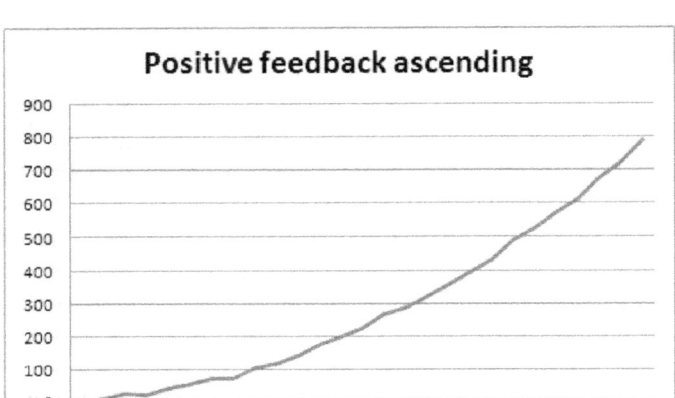

**Figure 22. Positive feedback ascending**

Examples of First-Order Dynamic Systems with positive feedback are: A savings account at a bank, with only retreats, a whale population where the percentage of births in each time period is high and deaths are rare.

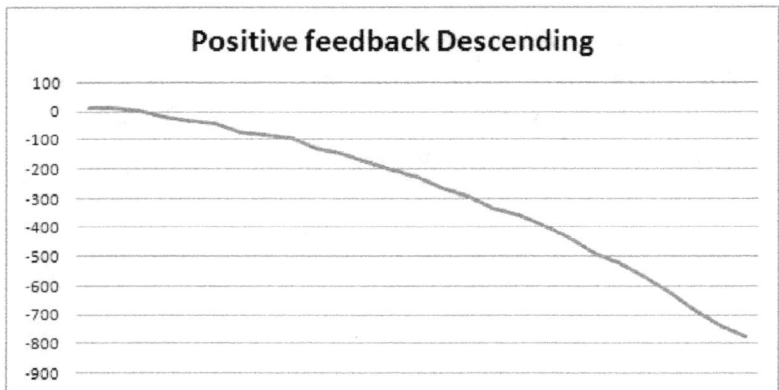

**Figure 23. Positive feedback Descending**

## Second-order dynamic systems

Second-order Dynamic Systems are those with two levels in its structure, as shown in Figure 24. Another way that presents the second-order systems if given auxiliary systems. In the second-order dynamical systems can be a feedback relationship, where both levels directly influence (See Figures 25 and 26)

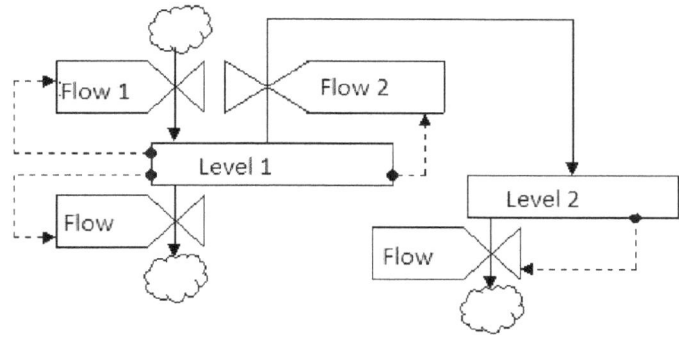

**Figure 24. Forrester diagram of a Second-order Dynamic System**

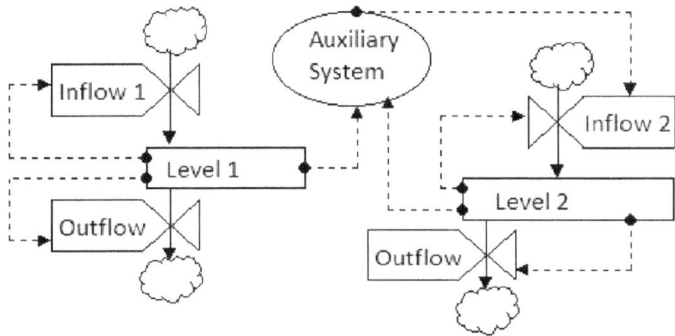

**Figure 25. Second-order Dynamic System with Auxiliary Systems**

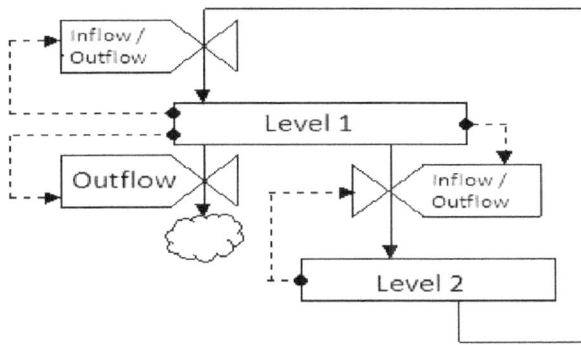

**Figure 26. Second-order Dynamic System with Feedback**

## PROBLEMS

1. A beer distributor buys from the factory 120 to 1800 boxes a day. Weekly sell 700 to 1200 boxes of cellar on the premises, and trucks from 1000 to 2500. In the month employees are rewarded with 5 boxes for your enjoyment if that month has not made "chicha" (breaking bottles accidentally) which if given is in the order of 1 or 2 boxes. Using System Dynamics to create the Synergistic diagram and the Forrester diagram.

2. The "parts-and-cia" company buying motherboards for computer directly from the manufacturer in the order of 200 to 1300 a month. The cards are stored in a cellar, which when it rains in the city, increasing its moisture, resulting in 75% of the time, its damage of 1 to 3% of the cards stored there. It is known that the probability of rain on the city during the month is 23.45%. "parts-and-cia" sold to "Tasra.com" from 24 to 93% of its inventory, of which from 0 to 2% leave with defects that are replaced with others. Tasra.com uses between 80 to 90% of your cards to assemble computers, where was damaged between 1 to 3% of those cards, which are used for parts or to be trash. Using System Dynamics to create the Synergistic Diagram and the Forrester Diagram.

3. In a farm where corn is grown there are two crops, the first named "Azucena" and the second called "Milk". In the "Azucena crop" is harvested per day between 600 to 1350 bags, and in the "Milk crop" between 100 to 256 bags. "Azucena crop" is sold between 70 to 98% of the total of sacks, while in the "Milk crop" is sold between 80 to 85% of was sold in the "Azucena crop". Using System Dynamics to create the diagram Synergistic and diagram Forrester.

4. "Rhatax.Com" sold complete computers and hard drives. Monthly purchase 10 to 20 computers and sold from 5 to 30 computers. As regards hard drives sold between 15 to 34 and buy 30 to 40, if the computers bought in the month does not exceed 12, and buy 40 to 55 hard drives, otherwise. Using System Dynamics to find stocks of computers and hard drives for a given number of months.

5. In a population, a virus called "politios infectus", infects among 10 to 14 people per hour. Health officials are alarmed because they die every 60 minutes, between 10% to 12% of those infected. Have developed a vaccine that is able to heal about 5 people per hour. Using System Dynamics to know how many hours are needed to cure an entire population

# COMPUTER MODELING FROM SYSTEM DYNAMICS

## Chapter 4

## COMPUTER MODEL STRUCTURE

### Structured programming

The **computer model** is constructed from diagrams Synergistic analysis of Forrester Diagrams and mathematical model is extremely simple:

Level subsystems and subsystems flow are represented by Level Variables and Flow Variables respectively. Similarly, the quantities are represented by random variables that store quantities of random distributions (most used is the Uniform distribution), the constants by variables constant, and other systems and subsystems for variables that store values.

In simple words, all the subsystems of Forrester diagram become variables and their types differ in the mode of declaration, assignment and corresponding operation. That is, a level variable in the computer model would store quantities like the Forrester diagram and a inflow variable contains the quantities will be added to a level variable.

### Object-oriented programming

In contrast, when building Object Oriented Computer Models take each of the types of subsystems Forrester Diagram to create different objects. That is, create objects level type, inflow type, outflow type, random type, etc., respecting their respective properties reflected in their methods. You need to create an inclusive environment for objects.

Can be summarized in a single object, the level subsystems and flow subsystems associated with it, where the functionality storage of levels, attributes would be carried out, and the modifying action of Flows, Methods would be carried out.

## TOOLS

## Programming language

The programming languages used in this text are the C++ and Java. Both C++ and Java are languages object-oriented programming used by millions of people around the world to develop applications on any operating platform, for that reason they were chosen.

Simulations of the case studies are conducted, in part, purely in sequential comands within a program written in C++ or Java, and in part to the use of DYNAMIC_SYSTEMS Class.

## DYNAMIC_SYSTEMS class in C++[1]

DYNAMIC_SYSTEMS Class is a purpose-built to perform simulations of Dynamic Systems based on the facilities of the Object Oriented Programming. This class is completely at Forrester diagram.

The formal definition of the class is as follows:
```cpp
//The file name is SistemasD.h
#ifndef DYNAMIC_SYSTEMS_H
#define DYNAMIC_SYSTEMS_H
#include <iotream.h>
#include <stdlib.h>
#include <conio.h>
class DYNAMIC_SYSTEMS {
 private:
        typedef Structure_levels {
              char *name;
              double value;
        } Levels[100];

        typedef Structure_flows {
              char *name;
              int Type; // 1- In 2- out 3 -In/Out 4- with Auxiliary
              int TipoInter; // 1- Normal 2- percentage
              int In_Level;
              int Out_Level;
              double Limits_Minimum_Change;
              double Limits_Maximum_Change;
```

---

[1] As reference of programming in C and C++ were used Cohoon, 1998; Jayanes, 1998, Main & Savitch, 2001; and Schidlt, 1995

```
        } Flows[500];

    typedef Structure_Auxiliary {
        char *name;
        double Limits_Minimum_Change;
        double Limits_Maximum_Change;
    } Auxiliary [500];

    int Numbers_Levels;
    int Numbers_Flows;
    int Numbers_Aux;
    int Numbers_Periods;

public: // Public elements
    DYNAMIC_SYSTEMS();
    ~DYNAMIC_SYSTEMS();
    void Create_Level(char *name, double InitialValue);
    void Create_Auxiliar(char *name, double Limits_Min,double Limits_Maximum);
    void Create_InFlow(char *name, char *Level,char *Name, double Limits_min,
        double Limits_Maximum, int Type_I);
    void Create_OutFlow(char *name, char * Name_Level,double Limits_Minimum,
        double Limits_Maximum, int Type_I);
    void Create_Flow_In_Out(char *name, char *Name_Level_1, char Name_Level_2,
        double Limits_Minimo,double limit_Max,int Type_I);
    void Create_Flow_In_Aux(char *name, char *Neme_Level,char *Names_Auxiliary);
    void Create_Flow_Out_Aux(char *name, char *Neme_Level,char *Names_Auxiliary);
    void Simulation(int Periods);
}; // end of the class
#endif
```

## PROBLEM OF THE CHICKEN POPULATION

**Statement:** We will base the case of the chicken population in chapter 3.
**Target System:** The objective is to calculate the monthly chicken population.
**Synergistic Diagram:** See Figure 16.
**Forrester Diagram:** See Figure 27

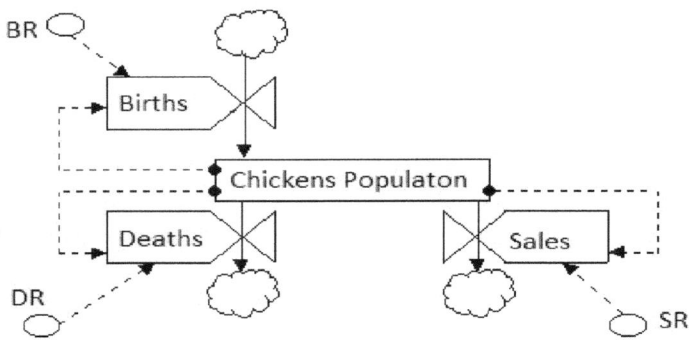

**Figure 27. Forrester diagram of the problem of Chickens**

## Mathematical model

We analyze the following aspects of Forrester diagram:
1. As we can see the chickens population represents a level subsystem, as well as being our object of interest, what accumulates a physical quantity as are the chickens.
2. The Births vary the amount of chickens in population level, therefore is classified as a flow, and its tendency to increase is said to have an inflow.
3. In the same way, sales and deaths vary the level of population decline, which is why they are categorized as Outflow.
4. The random variables BR(birth rate), DR (death rate) and SR(sales rate), represent events or products in the super-systems.
5. Applying the mathematical model of Forrester diagrams, we can say that the population of chickens is equals: is the same: to which is, more births, fewer deaths and fewer sales are made in the month.

Here, we describe the mathematical model of problem

$$PCH = PCH + BCH - DCH - CHS$$

Where,
PCH: POPULATION OF CHICKENS
BCH: BIRTH OF CHICKENS
DCH: DEATHS OF CHICKENS
CHS: CHICKEN SOLD

## Computer model based on C++

Is presented below a list of software written in C++, with which you can perform the computer simulation of the dynamic system described above.

```cpp
// name of the Program SPollos.Cpp
// used bookstores
#include <iostream.h>
#include <stdlib.h>
#include <conio.h>

// begin of the main program
int main () {
    // declaration of variables
        int PCH ,BCH ,DCH, months, i,CHS;
    // initialize variables with conditions initials
    PCH = 240; BCH = 0; DCH = 0;
    clrsrc(); randomize();

    // request how many months he/she wants to make the simulation
    cout<<"population's of Chickens Simulation \n";
    cout<<"With How many months he/she wants to make the Simulation";
    cin>>months;
    clrscr();
    // calculate of the population per months
    for(i=1; i <= months; i++) {
        // calculate the sales in the month
            CHS = PCH*random(10-0+1)/100;

        // calculate the births
            BCH = random(50-2+1)+2;

        // calculate the deaths
            DCH = random(5-1+1)+1;
          // to calculate the population of the month
            PCH = PCH + BCH - DCH - CHS;

            // present for screen the current population
              cout<<"The population of Chickens in the month"<<i<<"="<<PCH<<"\n";
    } // end of the for
    return 0;
} // end of the program
```

## Computer model based on Java

```java
// It name of the Program JSPollos.java
import java.io.*;
import javax.swing.*;
class JSPollos {
  public static void main(String [] args) {
        // declaration of variables
          String L;
          int PCH ,BCH ,DCH, months, i,CHS;
        // initialize variables
          PCH = 240;   BCH = 0; DCH = 0; CHS = 0;
        // to request the months of the simulation
          clrscr();
        L = "population's of Chickens Simulation \n";
        L = L + "With how many months you want to make the Simulation";
        L = JOptionPane.showInputDialog(L);
         months = Integer.parseInt(L);
        // calculate of the population per months
            for(i = 1; i <= months; i++) {
                // calculate the sales in the month
                   CHS = PCH * (int)(((10-0)*Math.random()+0)/100);

                // calculate the births
                  BCH = (int)((50-2)*Math.random()+2);

                // calculate the deaths
                   DCH = (int)((5-1)*Math.random()+1);

                // to calculate the population of the month
                   PCH = PCH + BCH - DCH - CHS;
           } // end of the for
         L = "The population of Chickens in the month "+(i-1)+" is "+PCH;
         JOptionPane.showMessageDialog(null,L);
        System.exit(0);
   } // end main
} // end of the class
```

# PROBLEM OF THE POPULATION OF ADULT RABBITS AND YOUNG RABBITS

## Statement

A Farmer is dedicated to raising rabbits. Buy 240 Conejos suitable for reproduction. For the experience of more than 20 years has been in business, you know that every month are born between 2 and 45 young rabbits that are not suitable for reproduction, but they will when they meet the month or so. In addition, die between 0 and 5 adult rabbits and only between 0 and 2 young rabbits. Every month the farmer sells from 0 to 10% of the adult population, and zero of the young population, since the government does not allow you to sell to young rabbits.

Were asks to perform the corresponding Synergistic diagram, Forrester diagram and a computer simulation of the studio system, showing the population of adult rabbits monthly for a number of months given.

## Objective

The objective is to estimate the population of adult rabbits monthly for a number of months since.

## Synergistic diagram

In the Synergistic diagram, takes into account the populations of both adult and young rabbits, deaths, births and sales. Figure 28 describes the relationships between these subsystems.

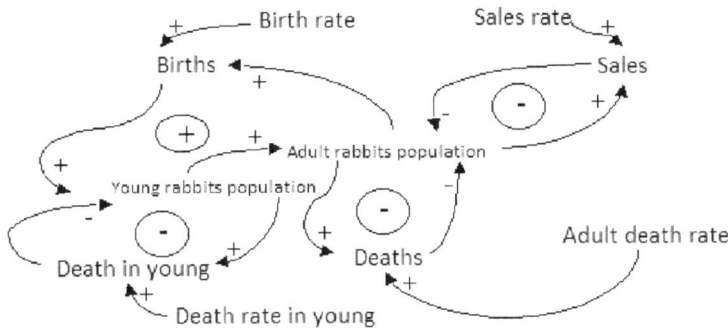

**Figure 28. Synergistic diagram of the case of adult and young rabbits**

As shown in Figure 28, there is a direct relationship between the rate of births to the births, which in turn increases the population of young rabbits, which in turn also increase to the adult population as they grow. On the other hand, it appears that the deaths decrements in both populations, as well as the sales only decrements the adult population.

## Forrester diagram

As we analyze there are two levels: the first is the young rabbits and the other is made up of adult rabbits. The young rabbit population has an inflow which are the births, and two outlets: the deaths and that they become adults.

On the other hand the adult population is fed (inflow) for the outflow of young population that has reached the "age of majority ". This level also presents two outflows, which represents the deaths, and the other the sales. It also presents the following random variables TN (birth rate), TMJ (rate of deaths of young rabbits) and VAT (sales rate of adult rabbits), TMA (rate of deaths of adult rabbits). Figure 29 describes the diagram for Forrester.

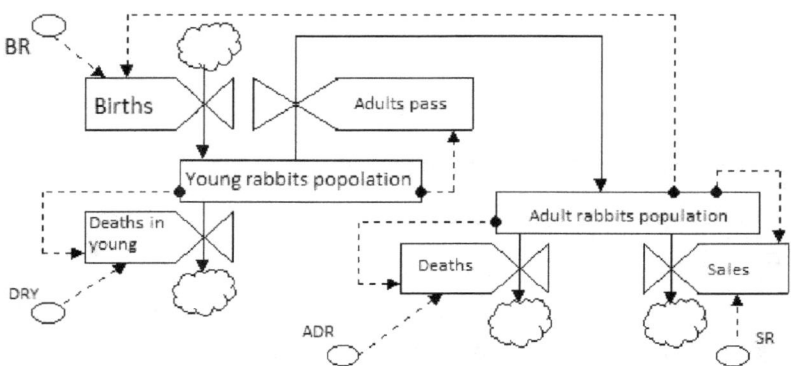

**Figure 29. Forrester diagram of rabbits**

## Mathematical model

From Forrester diagram (Figure 29) we have the adult population is equal to that which has more births a month ago, less the deaths and the sales carried out (Equation 1). Similarly, the population of young rabbits is equal to the births minus those who die young (Equation 2).

**(Equation 1)** PAR = PAR + RM1 −DAR - ARS
**(Equation 2)** PYR = BIR − DYR

Where,
**PAR**: POPULATION OF ADULT RABBITS
**RM1**: RABBITS FOR 1 MONTH
**DAR:** DEATHS IN ADULT RABBITS

**ARS**: ADULT RABBITS SOLD
**PYR**: YOUNG RABBIT POPULATION
**BIR**: BIRTHS
**DYR**: DEATHS IN YOUNG RABBITS

## Computer model based on C++

The simulation program is as follows:

```
// It names of the Program CONEJOS1.C

// Used bookstores
    #include <iostream.h>
    #include <stdlib.h>
    #include <conio.h>

int main () {
    // declaration of variables
        int PAR,RM1,DAR,ARS,BIR,DYR, months,PYR,i;

    // initialize variables with conditions initials
        PAR = 240;  RM1 =0; DAR = 0; ARS = 0; PYR = 0;

    // request how many months you want to make the simulation

        clrsrc(); randomize();
        cout<<" population's of Young Rabbits Simulation and Adultos\n";
        cout<<" With How many months you want to make the Simulation";
        cin>>months;
        clrscr();

    // calculate of the population per months

      for(i=1; i<= months; i++) {
        // to calculate the sales in the month
            ARS = PAR * random(10-0+1)/100;

        // to calculate the birth
            BIR = random(45-2+1)+2;
        // to calculate the deaths adults
            DAR = random(5-0+1)+0;
```

```
        // to calculate the young deaths
            DYR = random(2-0+1)+0;

        // to calculate the MATURE population of the month
            PAR = PAR + RM1 - DAR - ARS;

            if(PAR <0) PAR = 0;

        // to calculate the young population of the month
            PYR = BIR - DYR;
            if(PYR <0) PYR = 0;

            RM1 = PYR;

         // to present for screen the current population
            cout<<"Population of Mature Rabbits in the month";
            cout<< i << "it is "<< PAR << " \n ";
    } // end of the for

    return 0;

} // main end
```

## Computer model based on the class DYNAMIC_SYSTEMS

The model is as follows:

```
// It names of the Program RabbitsD.cpp
#include <iostream.h>
#include <conio.h>
#include "SistemaD.h"

// I begin it programs main
    void main () {
        // declaration of variables
        DYNAMIC_SYSTEMS Rabbits;
        int months;
        clrsrc();
        // request how many months you want to make the simulation
        cout<<"population's of Rabbits\n Simulation";
        cout<<"the months of the Simulation";
```

```
        cin>>months;

        // create the Levels
        Rabbits.Create_Level("Population of Adults Rabbits", 240);
        Rabbits.Create_Level("Population of Young Rabbits", 0);

        // create the Flows
        Rabbits.Create_Inflow(" Births ", "Population Young Rabbits", 2, 45,1);
        Rabbits.Create_OutFlow(Young Deaths", "Population Young Rabbits", 0, 2,1);
        Rabbits.Create_OutFlow(" Sales ", "Population of Adults Rabbits", 0, 10, 2);
        Rabbits.Create_OutFlow(" Deaths ", "Population of Adults Rabbits", 0, 5, 1);
     Rabbits.Create_Flow_In_Out("Adults Happen", Population of Adults Rabbits",
"Population Young Rabbits", 100, 100, 2);

        // carry out the Simulation
            Rabbits.Simulation(months);
 } // End of the Program
```

## Java-based computer model

The model is as follows:

```
// It names of the Program YRabbits1.java
import java.io.*;
import javax.swing.*;

public class YRabbits1 {
      public static void main(String [] args) {

       // declaration of variables
         String L;
         int i,PAR,RM1,DAR,ARS,BIR,DYR, MONTHS,PYR;

       // initialize variables
          PAR = 240;  RM1 = 0; DAR = 0; ARS = 0;  PYR = 0;

       // months of the simulation
          JOptionPane.showMessageDialog(null, "population's of Rabbits Simulation 2°");
              L = JOptionPane.showInputDialog ("With How many months wants to make
the Simulation");
             MONTHS = Integer.parseInt(L);
```

```
      // I calculate of the population per months
     for(i=1; i <= MONTHS; i++) {
       // to calculate the sales in the month
           ARS = PAR * (int)(((10-0)*MATH.RANDOM ()+0)/100);

       // calculate the Birth
           BIR = (INT)((50-2)*Math.random()+2);

       // calculate the deaths adults
           DAR = (INT)((5-0)*Math.random()+0);

        // calculate the young deaths
           DYR = (INT)((2-0)*Math.random()+0);

       // calculate the mature population of the month
           PAR = PAR + RM1 - DAR - ARS;
           if(PAR <0) PAR = 0;

        // calculate the young population of the month
           PYR = BIR - DYR;
           if(PYR <0) PYR = 0;

           RM1 = PYR;

         // present for screen the current population
          L = The population of Mature Rabbits in the month " +i+ " is +PAR);
          JOptionPane.showMessageDialog(null, L);

      } // end of the for
    }
} //fin of the class
```

## PROBLEM OF POOL OF WATER TO A TEMPERATURE

### Statement

In a refinery is needed for a particular critical process water is provides at a given temperature, such requests can vary over time. To satisfy these requests for water, there is a container in which it is supplies with both cold and hot water.

It is known that for every 10 liters of hot water entering in the container, the temperature rises one degree centigrade, and that for every 15 liters of cold water, also were manages to lower the temperature one degree centigrade. The special process demand water at a temperature between 50 to 80 ° C at a given time. Measurements are made every hour.

Do the corresponding Synergistic diagram, Forrester diagram and a computer simulation system, showing, the liters of cold water and hot water is needed to get the temperature. The initial temperature is 50 ° C.

## Objective

The aim is to calculate the liters of hot water and cold water are needed every hour to meet the critical process.

## Synergistic diagram

In the Synergistic diagram, the subsystems involved are water temperature, hot water, cold water, temperature critical processes. This describes in Figure 30.

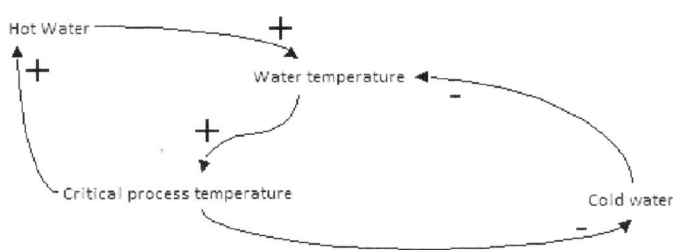

**Figure 30. Dada Synergistic diagram of a pond at given temperature**

## Forrester diagram

Based on the analysis of synergistic diagram (Figure 30) the problem that concerns us, find the following behaviors of the subsystems that comprise:

1. **Water Temperature:** It is a level subsystem, and stores the temperature of the water stored in the container.
2. **Cold Water:** is a inflow subsystem which is related to the level called Water Temperature. Since changes in temperature.
3. **Hot Water:** An inflow that is related to the level called water temperature.

4. **Critical Process Temperature:** Represents the information that comes from the super system and therefore is classified as an auxiliary system.

Figure 31 shows the Forrester diagram constructed from this analysis.

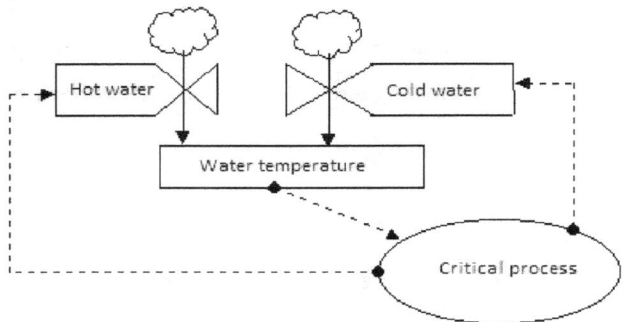

**Figure 31. Forrester Diagram**

## Mathematical model

The relationship between hot water and the vessel temperature is that by increasing the flow of hot water temperature rises. Similarly, the relationship between cold water and the temperature is that increasing the flow of cold water, the latter low.

In contrast, the relationship between water temperature and the critical process is based on how much the temperature satisfies the request. From this analysis, the mathematical model of this case study is described in the following equation:

$$TEMP = TEMP + (HOT/10) - (COLD/15)$$

Where:
TEMP: Temperature in degrees of water to be
HOT: Litres of hot water entering the vessel
COLD: Litres of cold water entering the vessel

The units of the equation are:
Grades = Grades + [Litres / (10 Litres / Grades) ] - [Litres / (15 Litres / Grades) ]

## C++-based computer model

The simulation program is as follows:

```cpp
// It names of the Program temp.cpp
// Used bookstores
    #include <iostream.h>
    #include <stdlib.h>
    #include <conio.h>

int main() {
    // declaration of variables
        int I, TEMP, HOT, COLD, HOURS, TPROCESS;

    // to initialize variables with conditions initials
        TEMP = 50;   HOT = 0; COLD = 0;

    // to request how many months he/she wants to make the simulation
        clrsrc(); randomize();

        cout<<" Simulation of the Pond of Water with a\n";
        cout<<" given Temperature \n";
        cout<<" With How many hours you want to make the Simulation";
        cin>>HOURS;

        clrscr();

    // calculate of the temperature per hour
        for(i=1;i <=HORAS;i++) {
         // calculate the temperature wants
            TPROCESS = RANDOM(80-50+1)+50;

         //  calculate if one needs water HOT or fry
            COLD = 0; HOT = 0;
            if(TEMP>TPROCESS) {// it dilutes cold
                COLD = (TEMP-TPROCESS)*15;
            }else {// HOT dilutes
                HOT = (TPROCESS-TEMP)*10;
            }

        TEMP = TEMP+HOT/10-COLD/15;

    // to present for screen the liters of water
        cout<<" * * * * * * * * HOUR" << I;
        cout<<" \nLos Liters of Water HOT that was used";
        cout<<" it was" << HOT << " \n ";
```

```
            cout<<" The Liters of Cold Water that it was used were";
            COut<<COLD<< " \n ";

    } // end of the for
    return 0;

  } // main end
} // end of the program
```

## Computer model based on the class DYNAMIC_SYSTEMS

The simulation program is as follows:

```
// It names of the Program TempSD.cpp
#include <iostream.h>
#include <conio.h>
#include "SistemaD.h"

void main() {
    // declaration of variables
        DYNAMIC_SYSTEMS Twater;
        int hours;

        clrsrc();

    // To request how many months he/she wants to make the simulation
        cout<<" Simulation of the Pond of Water with a";
        cout<<" given Temperature \n";
        cout<<" With How many hours you want to make the Simulation";
        cin>>hours;

    // create the Levels
        Twater.Create_Level("Temperature of the water", 50);

    // create the auxiliary variable
        Twater.Create_Auxiliar("Temperature of the process", 50, 80);

    // create the flows
        Twater.Create_Flow_In_Aux("COLD water", "Temperature of the water", "Temperature of the process");
        Twater. Create_Flow_In_Aux("HOT water", "Temperature of the water", "Temperature of the process");
```

```
    // To carry out the Simulation
        Twater.Simulation(hours);


} // End of the Program
```

## Java-based computer model

The simulation program is as follows:

```
// Name of the Program JTemp.java
import javax.swing.*;
public class JTemp{
     public static void main(String[] args) {
          // declaration of variables
            STRING L;
            int I, TEMP, HOT, COLD, HOURS, TPROCESS;

         // initialize variables
            TEMP = 50;  HOT = 0; COLD = 0;

      // request the months of the simulation
        L = Simulation of the Pond of Water with a given Temperature");
        JOptionPane.showMessageDialog(null,L);

        L = With How many hours you want to make the Simulation");
        L = JOptionPane.showInputDialog(L));

        HOURS = Integer.parseInt(L);
    // calculation of the temperature per hour
        for(i=1;i <=HOURS;i++) {
        // to calculate the temperature wants
            TPROCESS = (int)((80-50)*Math.random()+50);

        //calcular if one needs Hot water or Cold water
           COLD = 0; HOT = 0;
           if(TEMP>TPROCESS) {// cold
                COLD = (TEMP-TPROCESS)*15;
           }else {// Hot
                 HOT = (TPROCESS-TEMP)*10;
           }
           TEMP = TEMP+HOT/10-COLD/15;
```

```
                // present for screen the liters of water
                L = " HOUR " +i+"\n The Liters of Water HOT that was used was";
                L = L + HOT+ Liters \n The Liters of Cold Water that";
                L = L + it was used they were" + COLD+ " Liters ");
                JOptionPane.showMessageDialog(null,L);


        } // end of the for
    } // main end
} //fin of the class
```

## PROBLEMS

1. Building the mathematical model and computational model of problem 1 of Chapter 3

2. Building the mathematical model and computational model of problem 2 of Chapter 3

3. Building the mathematical model and computational model of problem 3 of Chapter 3

4. Building the mathematical model and computational model of problem 4 of Chapter 3

5. Building the mathematical model and computational model of problem 5 of Chapter 3

# CONCURRENT COMPUTER MODELING FROM SYSTEM DYNAMICS

Chapter 5

## INTRODUCTION TO CONCURRENT PROGRAMMING[1]

### Basic Definitions

*Concurrent Program*

A concurrent program is a program that has more of a logical line of execution, that is, a program that various parts of it are running simultaneously. A concurrent program can run on multiple processors simultaneously or not.

*Parallel program*

A parallel program is a concurrent program designed to run on parallel hardware.

*Distributed Program*

A distributed program is a parallel program designed to run on a network of autonomous processors that do not share memory.

*Process*

Process is called a running program with its associated environment. A process has several states: **running**, if the CPU is assigned, **ready or prepared**, if I could use the CPU if there is one available, **locked**, if an event waiting to happen before continuing execution.

*Concurrent processes or threads*

Concurrent processes are threads that generates a process of a concurrent program.

*Critical Section*

The **Critical Section** is the region of concurrent processing code, where made his most important task, which will have exclusive access to resources and / or data sharing, and other threads that need access, remain on hold.

---

[1] To confront the exposed thing with Lea, 2001 and Carretero, 2001

While a thread is in its critical section, others can continue its execution beyond their critical sections. When a thread leaves the critical section, then it should be allowed to proceed to other threads waiting to enter its own critical section, where there was a waiting process. Critical sections must be executed as quickly as possible and any thread not should be locked in its critical section.

## Principles of Concurrency

When two or more threads arrive at the same time to run and there is a relationship **(synergy)** between them (or are subsystems of a system), is said to have presented a concurrency of processes. Synergy between two concurrent processes is based on cooperation to carry out certain work and the use of information and shared resources.

### *Communication between threads*
**Communication between threads** involves the exchange of information between them, either through an implicit message or through a special type of variables called shared variables that are accessible to the code of these threads. Therefore, the communication thread manages the execution of a concurrent process influences the execution of another.

### *Thread synchronization*
**Thread synchronization** is usually necessary to preserve the integrity of shared variables. Synchronization does not allow two threads to access shared resources and information shared at the time, but instead is creating an atmosphere for exchange of control information which ensures that a concurrent process the exclusive use of a resource in favor of integrity.

### *Threads Competition*
**Threads Competition** occurs when the thread requires exclusive use of a resource, such as when two competing processes use the same shared variable.

### *Cooperation between threads*
**Cooperation between threads** occurs when two or more working in various parts of the same problem using the Communication and Synchronization.

## Characteristics of concurrent processes

### *Indeterminism*
Unlike the tasks specified in a sequential program are total order, the tasks of a concurrent program are of a partial product of the uncertainty in the exact order of occurrence of certain events, ie there is an undetermined in the execution. In simple terms, if run a concurrent program repeatedly with the same input data may produce different results.

*Synergy Threads*

The manifestations of the Synergy between concurrent processes or threads we can identify with:

- Threads share resources and compete for their access.
- The threads communicate to exchange data.
- Resource Management: A thread wishing to use a shared resource must request the appeal, expected to acquire, use and then release it.
- Communication: Communication can be synchronous, when they need to synchronize threads to exchange data, or asynchronous, where the wire that supplies the data does not need to wait to pick up the receiver thread, since the left in a temporary buffer communication.

## Concurrency problems

*Violation of Mutual Exclusion*

**Violation of Mutual Exclusion** occurs when two threads run their critical sections while accessing the same resource, how it would cause unpredictable results. To ensure mutual exclusion have the following options:

- **Semaphores.** A semaphore is a counter variable that controls the entrance to the critical region. P or WAIT operations, and V or SIGNAL; control, respectively, input and output of the critical region.
- One solution is the process to sleep (SLEEP) when waiting for a certain event, and wake WAKEUP when the event occurs.

*Deadlock*

A Thread is in a state of deadlock if it is waiting for an event that will never happen. There are four conditions that can cause the deadlock:

- The wires need to keep certain unique resources while waiting for others.
- The threads need exclusive access to resources.
- Resources are not available from threads that are waiting.
- There is a circular chain of threads in which each has one or more of the resources needed by following the chain.

*Indefinite delay*

A thread is in indefinite delays while waiting to be assigned a resource but never assigned. For example the release of a resource or a shared variable.

# CONCURRENCY IN JAVA[1]

## Introduction

The Java language supports concurrent programming through the Thread class, which contains methods that create, control and use properly threads.

One way to create a thread is to build a subclass of the Thread class. Within this subclass must define its public method called run (). In this method you should put the critical section code. Should be created an instance of the subclass (using the new statement), and to execute the thread should make a call to start () method to execute the run () method.

## Thread Class Methods

- **Start Method:** Used to initiate execution of a thread body, which is defined in the run () method.
- **Stop Method:** Used to complete the implementation of a Thread. No matter what the thread is doing, is dead, disposed of its internal state and release the resources were used.
- **Method suspend:** temporarily suspends the execution of the Thread.
- **Method resume:** Used to resume the execution of a thread suspended.
- **Sleep Method:** Put the Thread waiting for the time specified in the parameter.

## CONCURRENT COMPUTER MODEL STRUCTURE

### Structured Concurrent Programming

The structure of Concurrent Computer Model in the structured programming paradigm is:

Level subsystems correspond to **shared variables**, while both inflows and outflows represent **concurrent processes**, with the ability to access levels they choose, in an appropriate form.

### Object-oriented programming

In the paradigm of Object Oriented Concurrent Programming takes into account the characteristics of the Structure of Computer Model for structured concurrent programming for designing a level subsystems as objects whose function is to manage the shared information and on the other hand, the flow subsystem remain thread with the same functions.

---

[1] Jaworsky, 1999; Deldel & Deitel, 1999

# Object construction[1]

### *Levels*

The Levels in Concurrent Computer Model, are built as objects that control the shared variables. The general structure of the building is shown below (using the Java language):

```
public class Name_Level {
   // Private variables of storage
      private long Value = 0;

   // contructor
   public Name_Level(long variable_share) {
      Value = variable_share
   } // end contructor

    // Modifier method of the Level
      public synchronized void setValue(long x) {
         Value = x;
      }

    // Method that he/she Obtains the State or Value of the Level
      public synchronized long getValue() {
          return Value;
      }

} // end of the class
```

### *Inflows*

Inflows in the model correspond to threads that access levels to increase their values. The following describes the general structure:

```
public class Name_Inflow extends Thread {

  // Object of Associate Level
     private Name_Level m;

  // Contructor
     public Name_Inflow(Name_Level n) {
```

---

[1] Ibid; Jayanes, 1998

```
            m = n;
    } // end contructor

// Body of Execution
    public void run () {

    // Local variables
       long v;

       while(true) {
          try {
          // the value is increased
              v = (long)(100000 +300000*Math.random());

          // it modifies the level
                 m.setValue(v + m.getValue());

          // the thread sleeps
                 sleep((long)(5000*Math.random ()));

           }catch(Exception e) {}

        } // end while

     } // end run
} // end of the class
```

### Outflows

As inflows, outflows in the model correspond to the threads, the difference lies in that the latter decreases the value of Level objects. General structure is as follows:

```
public class Neme_Outflow extends Thread {

  // Object of Associate Level
      private Name_Level m;

  // Contructor
      public Neme_Outflow (Name_Level n) {
           m = n;
      } // end contructor
```

```
// Body of Execution

    public void run () {

    // Local variables
       long v;

       while(true) {
          try {
               // the value is decreases
                 v = (long)(100000+300000*Math.random ());

              // it modifies the Level
                    m.setValor(m.getValor () -v);

              // the thread sleeps
                    sleep((long)(5000*Math.random ()));

          }catch(Exception e) {}

       } // end while

    } // end run
} // end of the class
```
Note: can add that the other subsystems are equivalent to those of non Concurrent Computer Model.

## PROBLEM OF SAVINGS ACCOUNT

### Statement

John wants to open a savings account at a bank in the city to our local customers will pay consigned the work of building Web-oriented software he develops them. Know in advance that his clients pay on items ranging from 100,000 to 400,000 at any time of day.

John has a habit of removing at least 50,000 and a maximum of 150,000, but if it does not balance the amount you need, not withdrawn, pending further appropriations. Note: Assume that the laws do not allow savers to earn interest.

### Purpose of the system

Calculate the balance of the savings account in real time (Online).

## Synergistic diagram

Synergistic diagram is as follows:

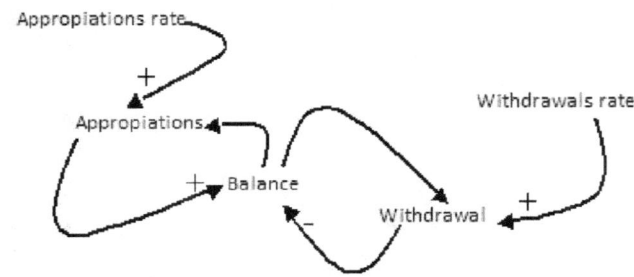

Figure 32. Synergistic diagram problem of savings account

## Forrester diagram

The Forrester diagram is a follows (Figure 33)

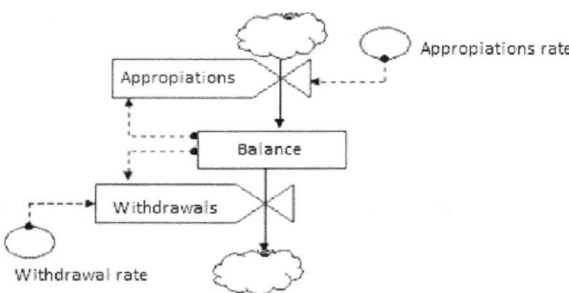

Figure 33. Forrester diagram problem of savings account

## Mathematical model

The Mathematical Mode is the following:

(1) Appropriations = 100000 +300000 * Math.random ()
(2) Withdrawals = 4000 * Math.random ()
(3) Balance = Balance + Appropriations - Withdrawals

## Computer model

The computer model consists of four (4) main types: The levels class , cong class  (Appropriations), Withdrawals class and class pal (Main).

**List of levels class**
```
import javax.swing.*;

class levels {
    private long Balance;
    private  DefaultListingModel mo;
    public levels(long s,  DefaultListModel no) {  Balance = s; mo = no; }
    public void setBalance(long x) { Balance = x; }
    public long getBalance() { return Balance; }
    public void setList(String x) { mo.addElement(x); }
}
```

**Listing of Appropiations class**
```
class Appropiations extends Thread{
      private levels m;
      public Appropiations (levels n){ m = n; }
      public void run(){
            long value;
       String s;
            while(true){
              try{
                  value = (long)(100000+300000*Math.random());
                  m.setBalance(valor + m.getBalance());
                  s = " It has been consigned: $"+value;
                  s = s + "Current balance is $"+m.getBalance();
                  m.setList(s);
                  sleep((long)(5000*Math.random()));
               }catch(Exception e){}
            }
      }
}
```

**Listing of Withdrawals class**
```
class Withdrawals extends Thread{
      private levels m;

      public Withdrawals (levels n){ m = n; }

      public void run(){
            long value;
```

```
        String s;
            while(true){
                try{
                    sleep((long)(4000*Math.random()));
                    valor = (long)(50000+100000*Math.random());
                        if(value > m.getBalance()){
                            s = " failed to withdraw from savings account: $";
                            s = s + valor +" Current balance is";
                            s = s + "$" + m.getBalance();
                        }else{
                                m.setBalance( m.getBalance()-valor);
                                s = " withdrew from the savings account: $"+valor;
                                s = s + " Current balance is $";
                                s = s + m.getBalance();
                        }
                        m.setList(s);


                }catch(Exception e){}
            }
        }
}
```

## Listing of Principal class

```
// Required packages
   import java.awt.event.*;
   import java.awt.*;
   import javax.swing.*;

public class Principal extends JFrame implements ActionListener {
// Buttons used
    private JButton Brun = new JButton("Run");
    private JButton BStop = new JButton("Stop");
    private JButton BExit = new JButton("Exit");
    private DefaultListModel model = new DefaultListModel();
    private JList list = new JList(model);
    private JScrollPane ps = new JScrollPane(list);
    private JPanel Row1 = new JPanel();
    private levels BALANCE;
    private Appropiations App;
    private Withdrawals WDRAWSLS;
```

```java
      private long value = 0;

  // Constructor without paramertros
      public Principal() {
         String s;
         s = "Concurrent Simulation Savings Account"
         setTitle(s)
         setSize(400.300)
         Container Cont = getContentPane();
         Cont.setLayout(new BorderLayout());
      // Placement of objects on the screen
         Row1.setLayout(new GridLayout(1,3));
         Row1.add(Run);
         Brun.addActionListener(this);
         Row1.add(Stop);
         Bstop.addActionListener(this);
         Row1.add(Exit);
         BExit.addActionListener(this);
         Cont.add(row1, "North");
         Cont.add(pd, "Center");
         Brun.setEnabled(true);
         Bstop.setEnabled(false);
         setVisible(true);
  } // End constructor

// Main Controller button actions
    public void actionPerformed(ActionEvent ae) {
        String s =(String)ae.getActionCommand();
        if(s.equals("Run")) {
           Brun_Click();
        else {}
        if(s.equals("Stop")) {
           BStop_Click();
        else {}
        BStop_Click();
        System.exit(0);}
     }

     } // End ActionPerf

// Execute Action Button
   private void Brun_Click() {
```

```java
        String s;
        s = JOptionPane.showInputDialog("Say the Initial Balance");
        value = Long.parseLong(s)
        BALANCE = new levels(VALUE, model);
        App = new Appropiations(BALANCE)
        WDRAWSLS = new Withdrawals(BALANCE)
        App.start();
        WDRAWSLS.start();
        Brun.setEnabled(false);
        BStop.setEnabled(true);


    } // End Brun
     private void BStop_Click() {
         App.stop();
         WDRAWSLS.stop();
         BStop.setEnabled(false);


     } // End BStop_Click

// Main Function
    public static void main(String[] arg) {

         Principal frm = new principal();

  } // End main
} // End of class
```

## PROBLEMS

1. Build the concurrent computational model for following problem: A car trader company, sold through its website cars of different brands. at a rate 20 to 33 family cars a month, from 3 to 17 trucks and 2 to 10 buses in the same time, the cars sold are delivered to customers three days after making the purchase. In the 20 branches it has in the country, sold in each(with a uniform distribution) at a rate of 10 to 15 family cars, from 5 to 10 trucks and from 5 to 7 buses per month.

2. Build the concurrent computational model following case: In the Systems Engineering Program at the XYZ University each semester of the following occurs:

    - Enter the first half from 25 to 85 new students.
    - For external transfer go from 0 to 10 students in the second to the seventh semester
    - In every semester, ie from first to tenth, lose half of 10 to 33% of the students.
    - In every semester, ie from first to tenth, out of the program is 0 to 2% of the students.

# COMPUTER MODELING CLIENT/SERVER FROM SYSTEM DYNAMICS

Chapter 6

## INTRODUCTION TO THE CLIENT SERVER PROGRAMMING

### Client Server model[1]

The Client / Server model describes the synergy between two processes running simultaneously. This model is based on a communication that consists of a series of questions and answers, an application begins execution and waits for the other will respond and then continues in this way the process. In this model there are two views:

**Customer Viewpoint:** The application that initiates the communication and is directed by the user. Customers usually perform functions such as:

- Handling the user interface.
- Capture and validation of input data.
- Generate queries and reports on the databases.

**Server Viewpoint:** The application that responds to customer requirements. These processes are running indefinitely. The servers have the following functions:

- Control concurrent access to shared resources.
- Links Communication with other networks.
- A customer service to request a corresponding server responds by providing it.

In designing the servers must include routines for proper management:

---

[1] Ortali et al, 1998

- **Privacy:** We must ensure that a user's private information, not accessed by an unauthorized party.
- **Authentication**: Verification of client identity
- **Authorization**: Make checks if a customer has access to the services provided by the server.
- **Data Security**: To these can not be accessed improperly.
- **Protection**: The system data and applications should not be monopolized.

Among the main features of the client / server architecture, are:

- The server changes involve little or no change in the client.
- The client need not know the logic of the server, only the external interface.
- The server presents to all its customers a unique and well defined interface.
- The customer does not depend on the physical location of the server, or the type of hardware which is, or your operating system.

## Essential infrastructure components Client / Server[1]

An infrastructure Client / Server consists of three essential components, namely:

**Operating Platform**. The platform must support all models Distribution Client / Server, all communication services, and will use industry standard components for distribution services.

**Application Development Environment.** A development environment should allow the coexistence applications for client and server processes developed by different programming languages or tools.

**Systems Management.** Although you can not avoid the system management functions(they increase considerably the cost of an IT solution and must fit the needs of the organization), are necessary to consider the following:

- What we need to manage?
- Where are they located processors and workstations?
- How many different types will be supported?
- What kind of support is needed and who provides it?

---

[1] To confront with Ibid and Tanambaum, 1992

## Advantages

**Increased productivity:**

- Users can use tools that are familiar, such as word processing, spreadsheets and tools to access databases.
- Through the integration of client / server applications with personal applications in common use, users can build particularized solutions that meet their changing needs.
- A consistent graphical user interface reduces training time applications.

**Lower operating costs:**

- Allows better utilization of existing systems, protecting the investment. For example, the sharing of servers(usually expensive) and peripheral devices(like printers) between client machines and allows for better overall performance.
- You can use components, both hardware and software from multiple vendors, which contributes significantly to reducing costs and enhances flexibility in the deployment and updating solutions.
- Provide better access to data. The user interface provides a uniform way of viewing the system, regardless of changes or updates to occur in it and the location of information.
- The movement of functions from a central computer or server to local clients, causes the displacement of the costs of the process toward smaller machines and therefore cheaper.

## Disadvantages

- There is a high technological complexity of having to integrate a wide variety of products.
- Requires a strong redesign of all the elements involved in information systems.
- It is more difficult to ensure a high degree of security in a network of clients and servers on a system with a single centralized computer. Checks should be made on the client and the server.

## JAVA CLIENT-SERVER PROGRAMMING

## Introduction to TCP/IP[1]

In this section we will explore one of the most interesting features of Java: the capacity for networking, TCP / IP. This allows you to use this language to build distributed applications in a very short time.

---

[1] To confront with Tanambaum, 1997

### IP addresses

All computers connected to an IP network, which uses the Internet Protocol are distinguished from one another by **IP address**. The IP address is a 32-bit number, usually expressed as 4 decimal numbers separated by dots. Each of these numbers corresponds to 8 bits of the IP address. For instance: 205.45.157.88. This IP address can be fixed or can be different each time the machine is connected to the network. This is what happens to almost all users who connect to Internet through phone line.

### Domain Names

These **domain names** are alphanumeric strings, easy to remember, that is associated with a single IP address. For example, Vulcano.com is the domain name associated with IP address 205.45.157.88. The Domain Name Server(DNS) are responsible for translating domain names into IP addresses. These servers maintain a table of correspondence between addresses and domains.

### Ports

The general way to communicate over the Internet is:

1. Specify the IP address of the machine you want to connect.
2. Specify the port number within that machine through which we communicate.

There are ports and to use presets(Standard RFC 1700):

- Port 80 HTTP server(Web server)
- port 21, FTP server
- Port 25 for SMTP(email)

## Communication via TCP[1]

### InetAddress Class

How to create an InetAddress object using the static method is InetAddress.getByName(String), which receives a host name in alphanumeric notation. For example, "www.tgs.com" or "204.45.157.88" and returns an InetAddress object to that direction. Furthermore, if the address does not exist or can not be found, this method throws an UnknownHostException. If you want to send packages to our own machine to use as a host name address "localhost" or "127.0.0.1 ". We may also use the method InetAddress.getLocalHost() which returns an InetAddress object that "points "to the local machine.

---

[1] Jaworsky, 1996

## Socket[1] Class

Java.net.Socket object is a "connector" through which send and receive data using TCP, and we do as we worked with a stream InputStream or OutputStream. Consider the following example that explains the workings of the Socket class:

**Example**

Assuming that a program is "listening" on port 1234 of the machine with IP address 209.41.57.70, the initialization of our Socket is:

```
InetAddress d = InetAddress.getByName("209.41.57.70");
Socket soc = new Socket(d,1234);
/ * Use the socket * /
...
/ * Close the socket * /
soc.close();
```

Once we have an open socket to another machine, we can obtain an inflow or outflow to receive or transmit data. This is done with the methods Socket.getInputStream() and Socket.getOutputStream():

Consider an example where we open a socket, we read the bytes transmitted from the other side and print on screen:

```
InetAddress d = InetAddress.getByName("209.41.57.70");
Socket soc = new Socket(d, 1234);
Inputstream is = soc.getInputStream();
while((int dato=is.read())!=-1){
   System.out.println("Recibido " + dato);
}
is.close();
soc.close();
```

## ServerSocket[2] Class

Java.net.ServerSocket class is the mechanism by which our programs may be "listening" in port waiting for incoming connections. The general form of working with sockets will be: a program we call "server" creates a ServerSocket on a port known for other programs.

---

[1] Deitel & Deitel, 1999
[2] Ibid

The server waits for a client to connect to it. At the time the connection is established, the two programs(client and server) get a Socket. By InputStream and OutputStream objects obtained through the Socket objects, the client and server exchange data. One of the two programs closes the connection. See how it performs the server side. The easiest way to create a ServerSocket object is indicating the port number to the constructor:

```
ServerSocket ssoc = new ServerSocket(1234);
/ * Use the ServerSocket object * /
```

Once created, we have to stay waiting for someone to attempt the connection. This is achieved by ServerSocket.accept() function. This function expects an incoming connection, and returns an object of type Socket.

```
ServerSocket ss = new ServerSocket(1234);
Socket soc = ssoc.accept();
/ * Use the Socket object * /
soc.close();
```

Once the server has the Socket, you can perform the same actions that the client, extract flows in / out, close the connection, etc. In the following example, our server waiting for incoming connection and responds with a welcome message:

```
ServerSocket servsock = new ServerSocket(1234);
Socket soc = servsock.accept();

OutputStream obs = s soc.getOutputStream();
String men = ";Hi!, how you doing?";
byte[] matrix = men.getBytes();

obs.write(matrix);
obs.close();
s soc.close();
```

ServerSocket an object can be obtained many different Socket objects, each independent of others. For example, we have a program that works on port 80 and assigns each new connection to a different execution thread. It is not necessary to close the previous Socket objects before accepting a new connection. This is what makes the Web server can accommodate several people at once, without waiting to finish with each client before the next meeting.

Suppose we have a class of objects "MiniServer", which implement the Runnable interface and are programmed to respond to requests that come to them through a Socket. Possible implementation for the server would be:

```
Serversocket ss = new ServerSocket(1234);
while(true){
   Socket s = ss.accept();
   MiniServer m = new MiniServer(s);
   Thread t = new Thread(m);
   t.start();
}
```

### Exception Handling[1]

The most common exceptions are:

- java.io.IOException. For cases where there is trouble connecting
- java.net.UnknownHostException. When specifying an IP address is unknown or incorrect.

## CLIENT-SERVER MODEL STRUCTURE

Preserving the entire structure of the previous computer models, client-server model is defined with the following configuration:

- The level subsystems are servers services
- Inflows and outflows correspond to Customers.

## PROBLEM OF ONLINE SAVINGS ACCOUNT

## Statement

John wants to open a savings account at a bank in the city to our local customers will pay consigned the work of building Web-oriented software he develops them.. Know in advance that his clients pay on items ranging from 100,000 to 400,000 at any time of day. John has a habit of removing at least 50,000 and a maximum of 150,000, but if it does not balance the amount you need, not withdrawn, pending further appropriations. Note: Assume that the laws do not allow savers to earn interest.

---

[1] Jaworsky, 1996

82 General Systems Theory: A focus on computer science engineering

## Purpose of the system

Calculate the balance of the savings account in real time(Online).

## Synergistic diagram

The Forrester diagram described in Figure 32

## Forrester diagram

The Forrester diagram described in Figure 33

## Mathematical model

The mathematical model contains the following equations:

**Client applications:**

(1) Appropriation = 100000 +300000 * Math.random()
(2) Withdrawals = 4000 * Math.random()

**Application Servers:**

(3) Balance = Balance + appropriation - Withdrawals

## Computer model

### Server

The server consists of the following items:

- **serverThread.** Server threads that manages the services to each customer.
- **server**. General Manager servers.
- **Message**. Shared object
- **Principal**. GUI server.

Listing of the serverThread[1] class

The list is as follows:

```
// Responds to customer
import java.net.*;
import java.io.*;
```

---

[1] All the classes were built with reference to programming in java, Ibid, Deitel & Deitel and Joyanes

```java
import javax.swing.*;

public class serverThread extends Thread {

  private Socket con;
  private String s;
  private Receive DataInputStream;
  private Send DataOutputStream;
  private Message m;

  public serverThread(Message N, Socket so) {
    try {
      m = n;
      con = so;
      Receive = new DataInputStream(con.getInputStream());
      Send = new DataOutputStream(con.getOutputStream());

    }Catch(Exception e) {
      m.setList("Error on Server. Connection" + con.getInetAddress());
      e.printStackTrace();
    }

  } // End construction

public void run() {
    long v;
    while(true) {
      try {
          s = Receive.readLine();
          if(s.equals("BALANCE")) {
             s = "Current Balance is:" + m.getBalance();
             SendMessage(s)
            m.setList("************************");
             s = "Balance sent to" + con.getInetAddress()+":$"+ s;
             m.setList(s)
             m.setList("************************");
          }else{
             v = Long.parseLong(s)
             if(v> 0) {
                  m.setBalance(m.getBalance() + v);
                  s = "appropriation of $" + v + "Completed. New Balance $";
                   s = s +  m.getBalance();
```

```
            SendMessage(s)

            m.setList("*************************");
            s = "Sent to" + con.getInetAddress();
            m.setList(s)
            s = "Appropriation $" + v + "Made";
            m.setList(s)
            s = "New Balance $" + m.getBalance();
            m.setList(s)
            m.setList("*************************");
        }else{
            if(v <0 & & abs(v) <= m.getBalance()){// Withdrawals
                m.setBalance(m.getBalance()-abs(v));
                s = "Removing $" + v + "Done"
                SendMessage(s)
                S = 'New Balance $ "+ m.getBalance();
                SendMessage(s)

                m.setList("*************************");
                s = "Sent to" + con.getInetAddress();
                m.setList(s)
                s = "Removing $" + v + "Done"
                m.setList(s)
                s = "New Balance $" + m.getBalance();
                m.setList(s)
                m.setList("*************************");
            }else{
                S = "Transaction Not done";
                SendMessage(s)
                m.setList("*************************");
                s = "Sent to" + con.getInetAddress()+":"+ s;
                m.setList(s)
                m.setList("*************************");
            }
        }
    }

    } Catch(Exception e) {
        m.setList("Server error message reading");
        e.printStackTrace();
    }
} // End while
```

```
    } // End run

    public void SendMessage(String s1) {
        try{
            Send.writeBytes(s1);
            Send.write(13);
            Send.write(10);
            Send.flush();
        }catch(Exception e){
            m.setList("Error Al Enviar Mensaje");
            e.printStackTrace();
        }
    }// fin SendMessage

public void Close() {
    try {
        con.close();
        m.setList("Server closed");
    }catch(Exception e){
        m.setList("Error closing");
        e.printStackTrace();
    }
  }// end close


 private long abs(long x)
 {
    if(x<0)x=-x;
    return x;
 }

} // end class
```

### Listing of the server class

The list is as follows:

```
import java.net.*;
import java.io.*;
import javax.swing.*;

class server extends Thread{
```

```java
private ServerSocket SERVER;
private String s;
private Message m;

public server(Message n) {
  m = n;
  try {
    m.setList("Wait ... Connecting the Server ..");
    SERVER = new ServerSocket(5000);
    m.setList( " Server IP Address:: "+InetAddress.getLocalHost());

    }catch(Exception e){
        m.setList("Server Error");
        e.printStackTrace();
    }
}// fin cons

public void run() {
    while(true){
      try{
          Socket con = SERVER.accept();
          (new serverThread(m,con)).start();

        }catch(Exception e){
            m.setList("Error reading message");
            e.printStackTrace();
      }

    }// end while

}// end run
} // end class
```

## Listing of the Message class

### The list is as follows:

```java
import javax.swing.*;

class Message {
    private  DefaultListModel mo;
    private long Balance;
```

```
      public Message(DefaultListModel no, long s) {
          mo = no;
          Balance = s;

      }
   public void setList(String x) {
          mo.addElement(x);
   }
   public void setBalance(long x) {
          Balance = x;
   }
   public long getBalance() {
          return Balance;
   }
} // end class
```

*Listing of the principal class*

The list is as follows:

```
   import java.awt.event.*;
   import java.awt.*;
   import javax.swing.*;

// Class Main extends JFrame
public class principal extends JFrame
       implements ActionListener{

// buttons used
   private JButton B1 = new JButton("Enable Server");
   private JButton B3 = new JButton("Exit");
   private DefaultListModel model = new DefaultListModel();
   private JList List = new JList(model);
   private JScrollPane pd = new JScrollPane(List);
   private JPanel Row1 = new JPanel();
   private servidor SERVER;

   private Message MEN;
   private long BALANCE;

   // Contructor
     public Principal() {
```

```java
        String s = " Server";
        setTitle(s);
        setSize(400,300);
        Container Cont = getContentPane();
        Cont.setLayout(new BorderLayout());

// Placement of objects on the screen

   Row1.setLayout(new GridLayout(1,2));
   Row1.add(B1);
   B1.addActionListener(this);
   Row1.add(B3);
   B3.addActionListener(this);

   Cont.add(Row1,"North");
   Cont.add(pd,"Center");
   setVisible(true);
 } // end constructor


 // Main Controller button actions
    public void actionPerformed(ActionEvent ae){

      String s = (String)ae.getActionCommand();

         if(s.equals("Enable Server")){
            B1_Click();
          } else {
              System.exit(0);}
} // end actionPerformed

// Action execute button
    private void B1_Click() {
      String s;
      BALANCE = 0;
      S = JOptionPane.showInputDialog("Say the Initial Balance");
      BALANCE = Long.parseLong(s);

      MEN = new Message(model,BALANCE);
      SERVER = new server(MEN);
      SERVER.start();
      B1.setEnabled(false);
```

```
        }   // End B1

// Main Function
    public static void main (String[] arg) {
        principal frm = new principal();

  } // End main
 } // End of class
```

## Appropriations Client

Appropriations client is composed of the following items:

- **Client.** Client application that requests the services.
- **Message.** Shared object
- **Principal.** Client GUI

*Listing of the Message class*

The list is as follows:

```
import javax.swing.*;

class Message {
    private  DefaultListModel mo;

       public Message(DefaultListModel no) {
        mo = no;
     }
      public void setList(String x) {
             mo.addElement(x);
     }
} // end of the class
```

*Listing of the client class*

The list is as follows:

```
import java.net.*;
import java.io.*;
import javax.swing.*;
class client extends Thread{
```

```java
private Socket con;
private String s;
private int res;
private DataInputStream Receive;
private DataOutputStream Send;
private Message m;

public client(Message n) {
  m = n;
  try {
    s = "Say The address of the server";
    s = JOptionPane.showInputDialog(s);

    m.setList("Wait ... Connecting the Client ..");
    con = new Socket(InetAddress.getByName(s),5000);
    m.setList("Clent connected to the IP Server: "+con.getInetAddress());

    Receive = new DataInputStream(con.getInputStream());
    Send =  new DataOutputStream(con.getOutputStream());

  }catch(Exception e){
     m.setList("Server Error");
     e.printStackTrace();
  }

}// end contructor

public void run() {

   while(true){
 try{
    s = Receive.readLine();
    if(s!=""){
        m.setList("Message Received: "+s);
    }// end if
  }catch(Exception e){
     m.setList("Error in the Client ");
     e.printStackTrace();
  }

   }// end while
}// end run
```

```java
    public void SendMessage() {
        try {
        s = " Say The Appropriation Amount ";
        s = JOptionPane.showInputDialog(s);
          Send.writeBytes(s);
           Send.write(13);
           Send.write(10);
            Send.flush();
            m.setList("Client sent appropriation of $'"+s+"'");

          } catch(Exception e){
             m.setList("Error sending message ");
             e.printStackTrace();
          }

      }// end SendMessage

public void Close() {
        try {
            con.close();
            m.setList("Server closed");
        }catch(Exception e){
            m.setList("Error closing");
            e.printStackTrace();
        }
    }// end close

} // end class
```

## Listing of the principal class

### The list is as follows:

```java
// Required packages
   import java.awt.event.*;
   import java.awt.*;
   import javax.swing.*;

// Class Main extends JFrame
   public class principal extends JFrame
        implements ActionListener{

   // Buttons used
```

```java
      private JButton B2 = new JButton("Client Action");
      private JButton B3 = new JButton("Exit");
      private DefaultListModel model = new DefaultListModel();
      private JList Lista = new JList(model);
      private JScrollPane pd = new JScrollPane(Lista);
      private JPanel Row1 = new JPanel();
      private cliente CLIENT;
      private Message MEN;

      public principal() {
         String s;
         S = "Client Appropiations";
         setTitle(s);
         setSize(400,300);
         Container Cont = getContentPane();
         Cont.setLayout(new BorderLayout());

         // Placement of objects on the screen
         Row1.setLayout(new GridLayout(1,2));
         Row1.add(B2);
         B2.addActionListener(this);
         Row1.add(B3);
         B3.addActionListener(this);

         Cont.add(Row1,"North");
         Cont.add(pd,"Center");
         setVisible(true);

         MEN = new Message(model);

         CLIENT = new client(MEN);
         CLIENT.start();


   } // end constructor

// Main Controller button actions
   public void actionPerformed(ActionEvent ae){

      String s =(String)ae.getActionCommand();

         if(s.equals("Client Action")){
```

```
                B1_Click();
            } else {
                System.exit(0);
            }

    } // end actionPerformed

// Execute Action

    private void B1_Click(){
        CLIENT.SendMessage();
    }// end B2

// Main Function
    public static void main (String[] arg) {
        principal frm = new principal();

    } // End main
} // End of class
```

## Withdrawals Client

Withdrawals client consists of the following items:

- **Client.** Client application that requests the services.
- **Message.** Shared object
- **Principal.** Client GUI

*Listing of the Message class*

The list is as follows:

```
import javax.swing.*;

class Message {
    private  DefaultListModel mo;
    public Message(DefaultListModel no) {
      mo = no;
    }
    public void setList(String x) {
          mo.addElement(x);
    }
} // fin de la clase
```

## Listing of the client class

**The list is as follows:**

```java
import java.net.*;
import java.io.*;
import javax.swing.*;
class client extends Thread{

  private Socket con;
  private String s;
  private int res;
  private DataInputStream Receive;
  private DataOutputStream Send;
  private Message m;

  public cliente(Message n) {
    m = n;
    try {
      s = "Say The address of the server";
      s = JOptionPane.showInputDialog(s);

      m.setList("Wait ... Connecting the Client ..");
      con = new Socket(InetAddress.getByName(s),5000);
      m.setList("Clent connected to the IP Server: "+con.getInetAddress());

      Receive = new DataInputStream(con.getInputStream());
      Send =  new DataOutputStream(con.getOutputStream());

    }catch(Exception e){
        m.setList("Server Error");
        e.printStackTrace();
    }


  }// end constructor

  public void run() {

     while(true){
    try{
       s = Receive.readLine();
      if(s!=""){
          m.setList("Message Received: "+s);
```

```
      }// end if
    }catch(Exception e){
       m.setList("Error in the Client ");
      e.printStackTrace();
     }
   }// end while
 }// end run

 public void SendMessage() {
       try {
        s = " Say the withdrawal amount ";
        s = JOptionPane.showInputDialog(s);
        Send.writeBytes(s);
        Send.write(13);
        Send.write(10);
        Send.flush();
        m.setList("Removing the client sent $'"+s+"'");

         } catch(Exception e){
             m.setList("Error sending message ");
             e.printStackTrace();
         }

    }// end SendMessage

 public void Close() {
        try {
            con.close();
            m.setList("Server closed");
         }catch(Exception e){
             m.setList("Error closing");
             e.printStackTrace();
         }
   }// end close

} // end class
```

## Listing of the principal class

The list is as follows:

```
// Required packages
   import java.awt.event.*;
   import java.awt.*;
```

```java
   import javax.swing.*;

// Class Main extends JFrame
   public class principal extends JFrame
         implements ActionListener{

   // Buttons used
      private JButton B2 = new JButton("Client Action");
      private JButton B3 = new JButton("Exit");
      private DefaultListModel model = new DefaultListModel();
      private JList Lista = new JList(model);
      private JScrollPane pd = new JScrollPane(Lista);
      private JPanel Row1 = new JPanel();
      private cliente CLIENT;
      private Message MEN;

      public principal() {
         String s;
         S = "Client withdrawals ";
         setTitle(s);
         setSize(400,300);
         Container Cont = getContentPane();
         Cont.setLayout(new BorderLayout());

         // Placement of objects on the screen
         Row1.setLayout(new GridLayout(1,2));
         Row1.add(B2);
         B2.addActionListener(this);
         Row1.add(B3);
         B3.addActionListener(this);

         Cont.add(Row1,"North");
         Cont.add(pd,"Center");
         setVisible(true);

         MEN = new Message(model);
         CLIENT = new client(MEN);
         CLIENT.start();

      } // end constructor

   // Main Controller button actions
```

```
    public void actionPerformed(ActionEvent ae){

      String s =(String)ae.getActionCommand();

         if(s.equals("Client Action")){
            B1_Click();
              } else {
                System.exit(0);
            }

      } // end actionPerformed

// Execute Action
    private void B1_Click(){
       CLIENT.SendMessage();
    }// end B2

// Main Function
    public static void main (String[] arg) {
         principal frm = new principal();

  } // End main
} // End of class
```

## Customer-Balance Class

The customer balances consists of the following items:

- **Client.** Client application that requests the services.
- **Message**. Shared object
- **Principal**. Client GUI

*Listing of the Message class*

The list is as follows:

```
import javax.swing.*;

class Message {
    private  DefaultListModel mo;
    public Message(DefaultListModel no) {mo=no;}
    public void setList(String x) { mo.addElement(x); }
} // end class
```

*Listing of the client class*

The list is as follows:

```java
import java.net.*;
import java.io.*;
import javax.swing.*;
class client extends Thread{

 private Socket con;
 private String s;
 private int res;
 private DataInputStream Receive;
 private DataOutputStream Send;
 private Message m;

 public cliente(Message n) {
   m = n;
   try {
     s = "Say The address of the server";
     s = JOptionPane.showInputDialog(s);

     m.setList("Wait ... Connecting the Client ..");
     con = new Socket(InetAddress.getByName(s),5000);
     m.setList("Clent connected to the IP Server: "+con.getInetAddress());

     Receive = new DataInputStream(con.getInputStream());
     Send =  new DataOutputStream(con.getOutputStream());

     }catch(Exception e){
         m.setList("Server Error");
         e.printStackTrace();
      }
  }// end constructor

public void run() {
    while(true){
       try{
           s = Receive.readLine();
          if(s!=""){
                 m.setList("Message Received: "+s);
            }// end if
         }catch(Exception e){
            m.setList("Error in the Client ");
```

```
                e.printStackTrace();
            }
    }// end while
}// end run

public void SendMessage() {
    try {
        s = "BALANCE";
        Send.writeBytes(s);
        Send.write(13);
        Send.write(10);
        Send.flush();
        m.setList("The client sends a query Balance");

        } catch(Exception e){
            m.setList("Error sending message ");
            e.printStackTrace();
        }

    }// end SendMessage

public void Close() {
        try {
            con.close();
            m.setList("Server closed");
        }catch(Exception e){
            m.setList("Error closing");
            e.printStackTrace();
        }
    }// end close

} // end class
```

## Listing of the principal class

**The list is as follows:**

```
// Required packages
    import java.awt.event.*;
    import java.awt.*;
    import javax.swing.*;

// Class principal extends JFrame
```

```java
public class principal extends JFrame
      implements ActionListener{

// Buttons used
   private JButton B2 = new JButton("Client Action");
   private JButton B3 = new JButton("Exit");
   private DefaultListModel model = new DefaultListModel();
   private JList Lista = new JList(model);
   private JScrollPane pd = new JScrollPane(Lista);
   private JPanel Row1 = new JPanel();
   private cliente CLIENT;
   private Message MEN;

   public principal() {
      String s;
      S = "Client Balance ";
      setTitle(s);
      setSize(400,300);
      Container Cont = getContentPane();
      Cont.setLayout(new BorderLayout());

      // Placement of objects on the screen
      Row1.setLayout(new GridLayout(1,2));
      Row1.add(B2);
      B2.addActionListener(this);
      Row1.add(B3);
      B3.addActionListener(this);

      Cont.add(Row1,"North");
      Cont.add(pd,"Center");
      setVisible(true);

      MEN = new Message(model);
      CLIENT = new client(MEN);
      CLIENT.start();

   } // end constructor

// Main Controller button actions
   public void actionPerformed(ActionEvent ae){

      String s =(String)ae.getActionCommand();
```

```java
            if(s.equals("Client Action")){
                B1_Click();
                } else {
                    System.exit(0);
                }

        } // end actionPerformed

// Execute Action

    private void B1_Click(){
        CLIENT.SendMessage();
        }// end B2

// Main Function
    public static void main (String[] arg) {
        principal frm = new principal();

    } // End main
} // End of class
```

## PROBLEMAS DE CAPITULO

1. From the dynamics of systems, building the client-server computer model of the following problem: Somewhere in the Milky Way whose name I remember, a spacecraft that found a planet with two civilizations that inhabited it. The planet's inhabitants belonged to different species of humanoids. The first civilization, which we call civilization "x" was the dominant civilization and had for ten years, a genetic disease that showed that individuals of this race die by physical exhaustion at a rate of between 100 to 200 per month and secondly, just born between 5 and 20 babies a month of this civilization, but its population was great to get the spacecraft, some 2,000,000 of "individuals x " across the globe.

   The other civilization which we call civilization "Y", was a small population of individuals of only 500,000. Had a birth rate of between 2 and 10 humanoids per month, with a death toll of between 0 and 4 per month.

   After two months of arrival at the planet, the vessel Diplomat Type J, a research scientist at the spacecraft discovered that civilization "Y" containing a gene that could cure the disease of "individuals-x". This gene into effect on "individuals-x" only three months after being injected. When vaccinated with the gene to "individuals-x", the genetic transformation such that the individual was in fact born a new race. In this race, which they called civilization "z", was characterized by the ability to procreate only with those of the same species and the "Y", ie, the "z" and "x" were inconsistent for procreate. The new race had a birth rate of between 1 and 5 deaths per month from 0 to 3 month. Note: and after that lived happily ever after?

2. Spaceship Endora in its mission of discovering new civilizations, finds a world Minshar class(M class) with a population of humanoids with less technological developments incapable of interstellar travel. Carrillo Captain down to earth with his team to study this civilization. Endora's team in his first Observations, were aware that there are two humanoid races: the Laakai and Naari.

   The Laakai are the race that dominates this society with a predominant order to have many elements subservient. We also find it impossible to cross between the two humanoid races, as Laakai have a genetic condition that does not. Interacting with a bit more Laakai, Carrillo captain discovers that the past two years many Laakai over 20 years began to suffer from a genetic disease that prevents them breed and die within days of expressed symptoms and signs of degeneration genetics. The population is 50 million Laakai the arrival of Endora. The following table describes the age distribution:

| Age | Pob | % | Age | Pob | % | Age | Pob | % |
|---|---|---|---|---|---|---|---|---|
| 0 | 2.500.000 | 5% | 8 | 1.000.000 | 2% | 16 | 1.000.000 | 2% |
| 1 | 2.500.000 | 5% | 9 | 500.000 | 1% | 17 | 2.500.000 | 5% |
| 2 | 1.500.000 | 3% | 10 | 1.000.000 | 2% | 18 | 2.500.000 | 5% |
| 3 | 1.000.000 | 2% | 11 | 500.000 | 1% | 19 | 3.000.000 | 6% |
| 4 | 1.000.000 | 2% | 12 | 1.000.000 | 2% | 20 | 4.000.000 | 8% |
| 5 | 500.000 | 1% | 13 | 500.000 | 1% | > 20 | 19.500.000 | 39% |
| 6 | 1.000.000 | 2% | 14 | 1.000.000 | 2% | | | |
| 7 | 1.000.000 | 2% | 15 | 1.000.000 | 2% | | | |

The births of the Laakai is about 0.5% to 2% per month(if not adults, no births). Deaths from natural causes are of the order, at all ages, from 7% to 11%. Month for deaths genetic defect found in the order of 80% to 97% of "sick. " Patients genetically(over 20) grew monthly rate of 1% to 12% of the population over 20 years. The deaths for month by cause the genetic abnormality found in the order of 80% to 97% of "sick". Patients(over 20) grow monthly rate of 1% to 12% of the population over 20 years.

Dr. Nicolas Xofh of the team, achieved on the basis of the blood of Naari, that a Laakai patient recovers his health, but a change of the health lose their reproductive capacity. The vaccine gives immunity for 5 years, which may then fall sick again and again to be vaccinated. The production of vaccines monthly, just enough to be applied to between 75% to 78% of patients. On the other hand, the initial population of 2 million Naari, their birth is about 200 to 345,000 monthly and natural deaths are the order of 6% to 19%. Using System Dynamics to perform a simulation for describing the characteristics of these civilizations

# *SYSTEM DYNAMICS TO UML*

## Chapter 7

## INTRODUCTION

UML, Unified Modeling Language, above all, is a language that uses a graphical representation to describe a system. Its main objectives are:

- Express in graphical form a system to allow other people can understand.
- Specify the characteristics of a system before its construction.
- Systems are built from models specified.
- Document the system.

UML began to gestate in late 1994 when Jim Rumbaugh was linked to the company Rotional of Mr. Grady Booch, where the main objective was to find a way to combine the Booch method and the method OMT(Object Modeling tool) and so define a standard notation in the processes of analysis and design of object-oriented software. Joined the project after Mr. Ivar Jacoson, among the three created the first version of UML, which was offered to the OMG[1]. The OMG proposed a number of changes reflected in version 1.1 of UML, which was accepted as standard in November 1997[2].

UML to be a formal modeling language provides the following advantages:

- The specification has more rigor
- It allows to verify and validate the model made
- You can automate processes
- Generate code from models and vice versa, ie from source to generate the models and provided the model and date code are

UML is a graphical modeling language diagrams using several of the most important are:

- Use Case Diagram
- Class Diagram
- Sequence Diagrams

---

[1] Object Management Group  http://www.omg.org
[2] Booch, G. Rumbaugh, J. Jacobson, I, 1999.

106 General Systems Theory: A focus on computer science engineering

- Collaboration Diagram
- State Diagram
- Activity Diagram
- Component Diagram
- Deployment Diagrams

## PHASE ANALYSIS IN UML

In this section will instruct the form to how to make the most important UML diagrams of the analysis phase, which can be generated through the system dynamics. The diagrams of the analysis phase to be taken into account are: the Conceptual Model, Use Case and Sequence.

## Conceptual model of the system

### *Introduction*

A Conceptual Model is a graphical representation of concepts in a problem domain. This model will contain the relevant concepts and associations between them. The representation of concepts is done by means of rectangles divided into two areas, the top where you put the name that identifies it and the lower one describes the attribute list. The associations are represented by lines on which you place the name and the multiplicity of the association. Figure 34 shows an example where it describes the Conceptual Model Icons.

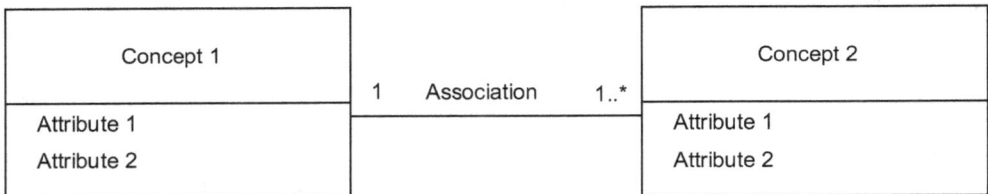

**Figure 34. Example of a Conceptual Model**

### *Building a conceptual model from the System Dynamics*

If we start, in the first place, the fact that the term concept refers to ideas, things, parts, objects, and why not to Systems. And second, the central objective of the creation of conceptual models is to decompose the problems in individual concepts and show their associations. And third, that the main function of Synergistic diagrams, is to describe the systems(concepts) involved in the problem, and their interactions(associations), can conclude that:

### Synergistic Diagram of the System Dynamics corresponds at the Conceptual Model of UML

That is, the subsystems of Diagram Synergistic correspond to the concepts and relationships to partnerships. Partnerships can describe or classify depending on the type of relationship or synergy, which present in synergistic diagram, namely:

1. *Positive relationship.* Positive relationships in a Synergistic diagram generates an association such as "Increase" where there is increased either abstract or real.

2. *Negative relationship.* Negative relationships Synergistic diagram generates an association of type "Decrease" where there is decreased either abstract or real.

3. *Neutra relationship.* Neutra relationship in a Synergetic diagram generates an association of type "Increase-Decrease" where there is an increase or decrease either abstract or real.

As an example, let us build the Conceptual Model of the Problem of Savings Account from its Synergistic Diagram depicted in Figure 35.

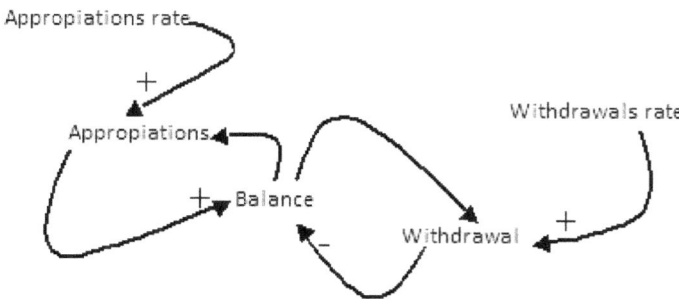

**Figure 35. Synergistic Diagram, Case Savings Account**

As can be seen, the subsystems of Diagram Synergistic Appropriation, withdrawals, balance, Appropriation rate and Withdrawals rate are taken as concepts and relationships as partnerships[1]. It should be noted that the associations are classified depending on the type of relationship that generated it. The explanation of the associations detailed in Table 2.

---

[1] To confront Figures 35 and 36

108 General Systems Theory: A focus on computer science engineering

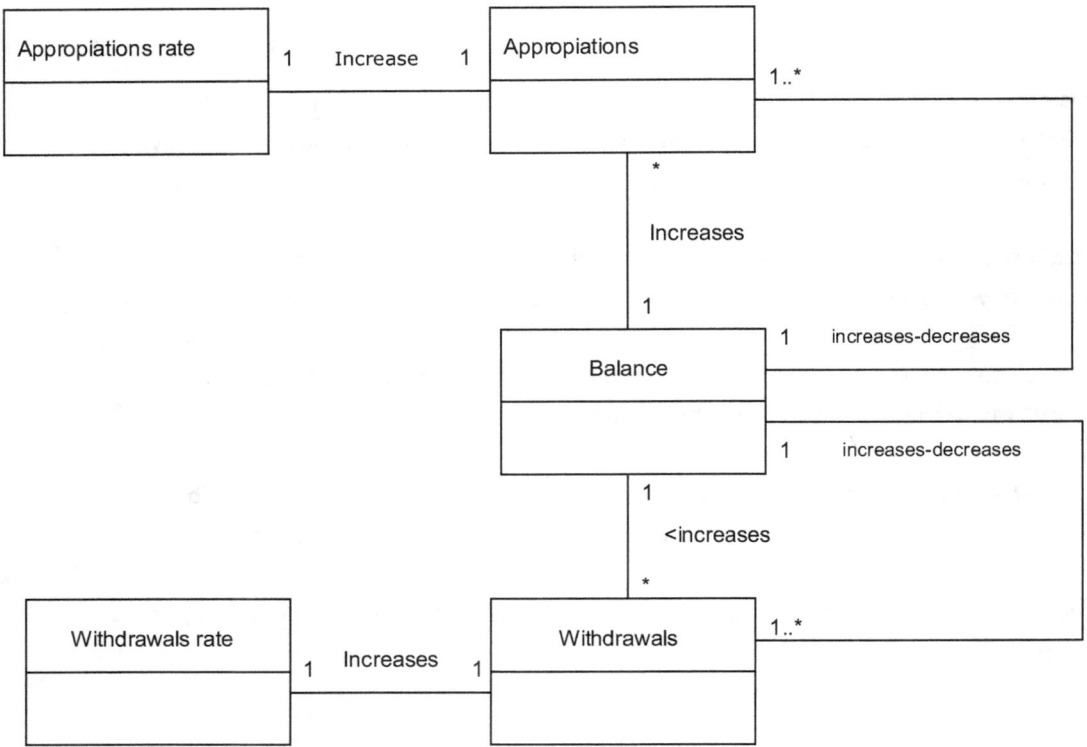

Figure 36. Conceptual Model of case savings account

Table 2. Details of the Associations of concepts

| Association | Explanation |
|---|---|
| Appropriation Rate Increases Appropriation | When the appropriation rate grows / decreases in a value, the same effect grows / decreases for the same value occurs in the Appropriations |
| Withdrawals Rate Increases Withdrawals | When the withdrawal rate grows / decreases in a value, the same effect grows / decreases occurs in the Withdrawals |
| Appropriation Increase Balance | Among more appropriations are made, the more it grows the Balance |
| Decreases Balance Withdrawals | Among more Withdrawals are made, the more decreases the balance |
| Balance Increases-Decreases Appropriation | By having a particular Balance(High or low) can not determine if motivated to do more or less appropriations, as the decision becomes subjective |
| Balance Increases-Decreases Withdrawals | By having a particular Balance(High or low) can not determine if motivated to do more or less Withdrawals, as the decision becomes subjective |

# Use cases

## Introduction

A Use Case is a document that describes a series of events made by an external agent at system, called the actor[1], its main objective the description of processes.

Use cases in UML are represented by intermediary of ovals, and these are associated with a name. The actors are usually represented by a human figure. The actors and use cases are associated with arrows directed. Figure 37 describes the icons of the Actors and Use Cases

**Figure 37. Use Case icon**

Below is the high-level format for describing use cases[2]:

| | |
|---|---|
| **Use Case:** | Use Case Name |
| **Actors:** | Actor1, Actor2,... |
| **Type:** | • **Primary**(important common processes), **Secondary**(minor processes or rare) or **optional**(may or may not be addressed) |
| | • **Essential**(expressed in theory and low-tech) or **Real**(Concrete: Subject to specific technologies in and out) |
| **Description:** | Description of process steps |

## Construction of the Use Cases from the System Dynamics

Use Cases are obtained from flows subsystems, and the levels subsystems, which are described in the diagrams of Forrester. The first, by modifier function, and last by their storage function. The Potential actors in this case will be the system itself and external users.

Taking for example, the Forrester diagram of the problem of savings account discussed in previous chapters(see Figure 38), find that it is composed of 2 flow subsystems one called Appropriations and

---

[1] Larman, 2002
[2] Larman, 1999

another called withdrawals, and a Level subsystem called Balance as well would have 3 cases of use that can be called **appropriation, withdrawals** and **Show Balance**, respectively.

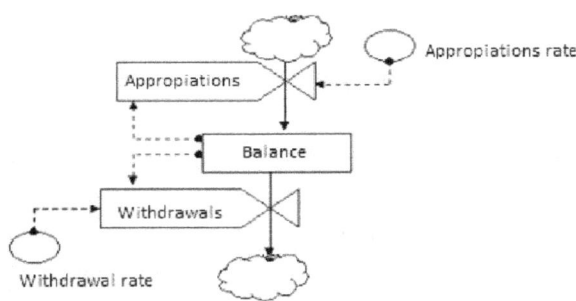

**Figure 38. Forrester diagram: Account Problem**

We also found that from the statement, the actors would be an external agent can be called user and the other actor would be the same system. The description of the use cases of high level are contained in Tables 3, 4 and 5, while Figure 39 describes the Use Case Diagram

**Table 3. Description of Appropriation Use Case**

| Use Case: | Appropriation |
|---|---|
| Actors: | User, System |
| Type: | Primary |
| Description: | The user arrives at the bank with a leaflet indicating the amount to be entered. System increases the balance. |

**Table 4. Description of withdrawals Use Case**

| Use Case: | Withdrawals |
|---|---|
| Actors: | User, System |
| Type: | Primary |
| Description: | The user arrives at the bank with a leaflet that tells how much to remove. The system verifies whether it can withdraw the amount, if you can decrease the balance. |

**Table 5. Description of Show Balance Use Case**

| Use Case: | Show Balance |
|---|---|
| Actors: | User, System |
| Type: | Primary |
| Description: | The user arrives at the bank to ask how much information is in the balance. The system verifies the quantity and shows the balance. |

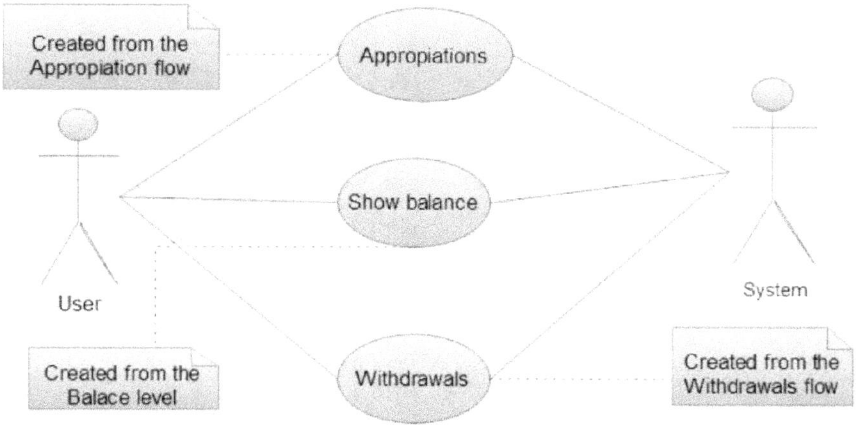

**Figure 39. Use Case Diagram of the Problem Savings Account**

# Sequence diagram of the analysis phase

## Introduction

Sequence Diagrams show in graphic form the events flowing from the actors at the system, with this one tries to analyze the behavior of the system[1]. In simple words, Sequence Diagrams, seek to describe in particular the course of events in a use case. It is important to note that in sequence diagrams are designed to describe which makes the system but not how[2]. The Sequence diagrams is made up of the following parties:

- An Actor
- An Instance of the system. Seen as Black Box
- Events. Input external event that occurs in the system Actor
- Operations. Action executed in response to a system event.

The graphical presentation of the components of sequence diagrams are:

- Actors are represented equally in the use cases, the system with a rectangle
- Events like arrows, associating its relevant information

In the Figure 40 describes the parts of a sequence diagram of the analysis phase.

---

[1] To confront with chapter 3 in the Section corresponding to the System Dynamics
[2] Confront with Larman, 1999 and with Larman, 2002

112 General Systems Theory: A focus on computer science engineering

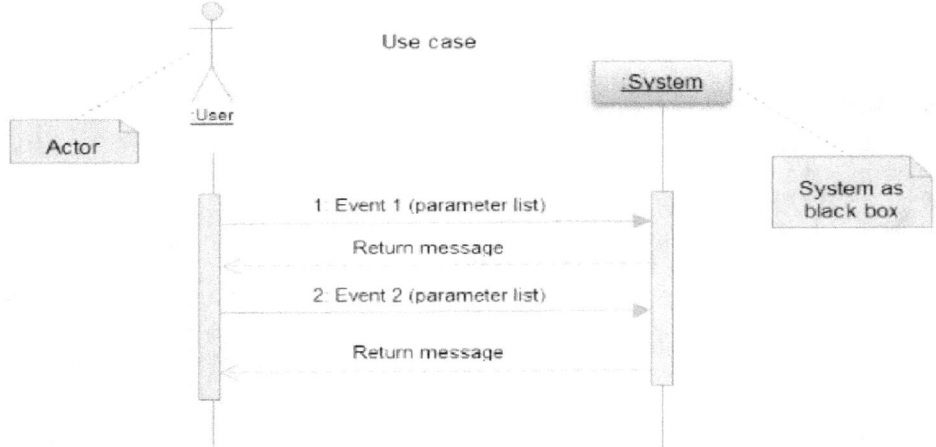

**Figure 40. Constituent parts of a sequence diagram of the analysis phase**

***Construction Sequence Diagrams of the analysis phase from the System Dynamics***

Known that the Use Cases are built in part by the flow subsystems, resulting what the sequence diagrams of the analysis phase for these use cases, present 3 basic events, such as:

1. The calculation of the variation in the level
2. Feasibility assessment to update
3. And the update itself.

And in the second instance, the use cases are built from the subsystemslevel, in which event occurs related to the action to show the value of the content it stores. Sequence Diagrams of the analysis phase in its general form, built from flow subsystems are described in Figure 41, lie to the general sequence diagram of use cases created from level subsystems are shown in Figure 42.

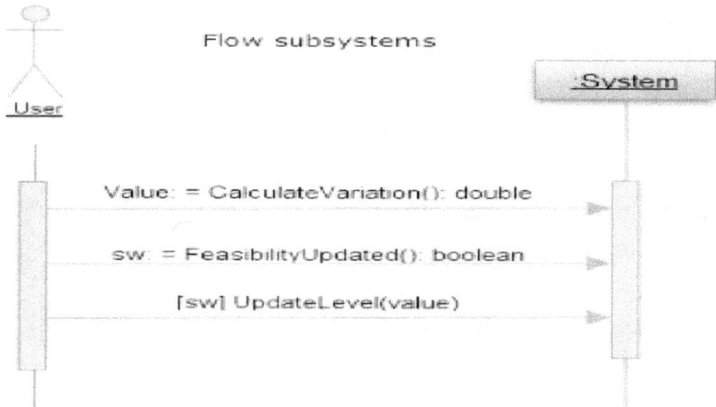

**Figure 41. General Sequence diagram generated from Flow Subsystems**

**Figure 42. General Sequence diagram generated from Level Subsystems**

As an illustration of the construction of sequence diagrams from system dynamics, it will build for the Use Cases Appropriations, withdrawals and Show Balance described above in Figure 39.

## *Appropriations Use Cases*

Appropriations Use Cases made two of the three general events usually do this type of use case built from the System Dynamics: CalculteVariationApp() and UpdateBalance(), since the banks usually do not place restrictions on the amount of that provision. The sequence diagram is depicted in Figure 43.

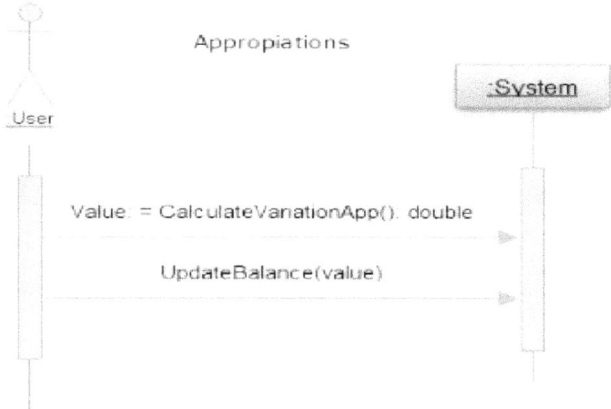

**Figure 43. Sequence Diagram of Appropriations Use Cases**

## *Withdrawals Use Case*

Withdrawals Use Case made three general events: CalculateVariationWithdrawals(), FeasibilityUpdated() and UpdateBalance(). The sequence diagram is depicted in Figure 44.

### Show Balance Use Case

Show Balance Use Case holds an event that bears the same name. The sequence diagram is depicted in Figure 45.

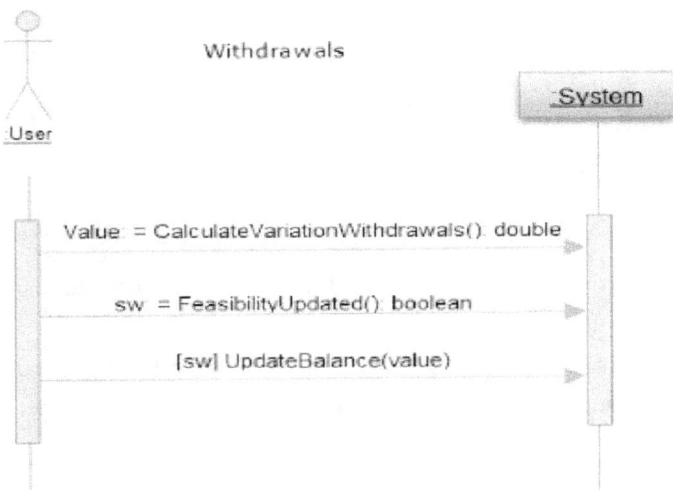

Figure 44. Sequence Diagram of withdrawals Use Case

Figure 45. Sequence Diagram of Show Balance Use Case

# UML DESIGN PHASE

This section describes how to construct the diagrams of the design phase, in UML from the system dynamics. Among the diagrams that are discussed in this section are: Sequence diagram of the design phase and the class diagram. In the creation of class diagrams is considered the type of programming such as: Object, Concurrent and Client - Server.

# Sequence diagram of the design phase

## Introduction

This section explains the sequence diagrams used for design. Which differ from those of the analysis phase diagrams that in the design phase describes the events that occur within the system.

## Building of Sequence Diagrams of the design phase from the System Dynamics

*Sequence diagrams of the design phase from use cases built based on level subsystems.* In the event ShowCurrentLevel() involving the user and the level, at which level returns at the system the actual amount saved. The system then prints the amount.(See Figure 46)

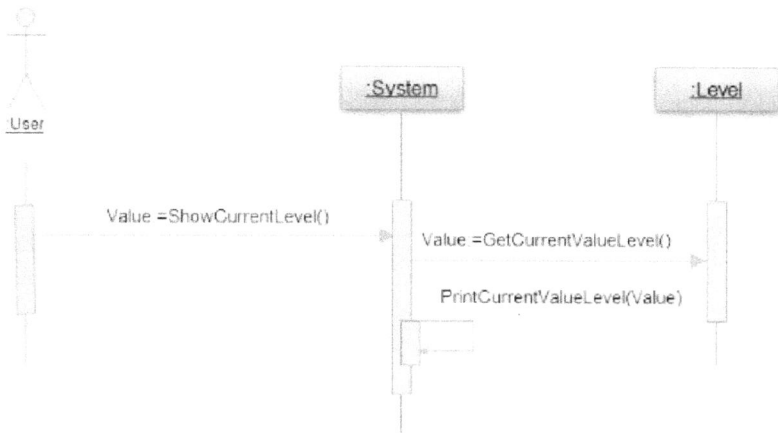

**Figure 46. Sequence diagram of the design constructed from subsystems level**

*Sequence diagrams of the design from use cases built based on flow subsystems.* We have the sequence diagram of the analysis phase, describes three events(CalculateVariation, FeasibilityUpdated and UpdateLevel), the idea is to describe how the system works "inside" each of these events:

In the event **CalculateVariation**, involving the user, the level associated with the flow and rate of change, where the latter returns at the a value system, generated under a specific policy[1], which corresponds to the rate of change. The system in turn prompted for the level you need it, the amount stored, and then calculate how much to vary the level.

---

[1] Between the used policies more we found the distributions random

In the event **FeasibilityUpdated,** involving the user and the level associated with the flow, where the level returned to the system a true value if the requested update can be performed, or the value of false otherwise. Likewise, the system returns this value at the user.

In the event **UpdateLevel**, involving the user and the level. Here the level updates the quantity it contains.

In the Figure 47 describes the sequence diagram of the design phase of a use case built through flow subsystems.

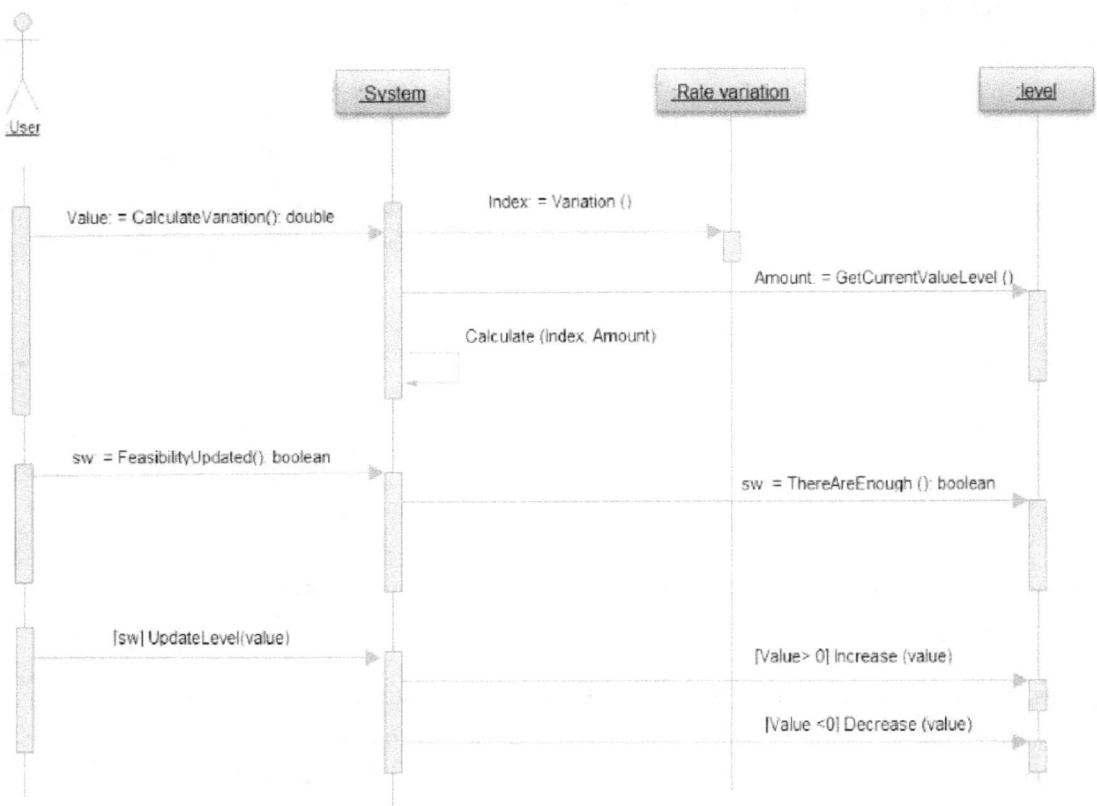

**Figure 47. Sequence diagram of the design phase of a use case built through flow subsystems.**

By way of example described in Figures 48, 49 and 50 the design sequence diagrams of use cases appropriations, withdrawals and ShowBalance respectively.

**Figure 48. Sequence diagram of the design phase of the appropriations use case**

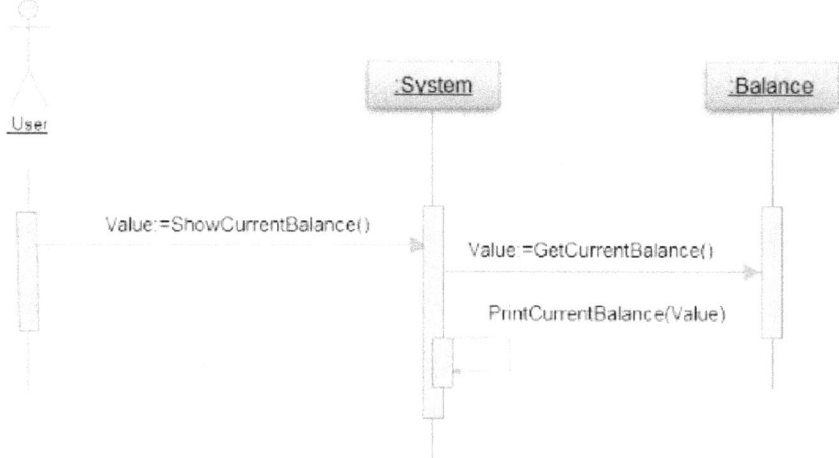

**Figure 49. Sequence diagram of the design phase of the MostrarSaldo use case**

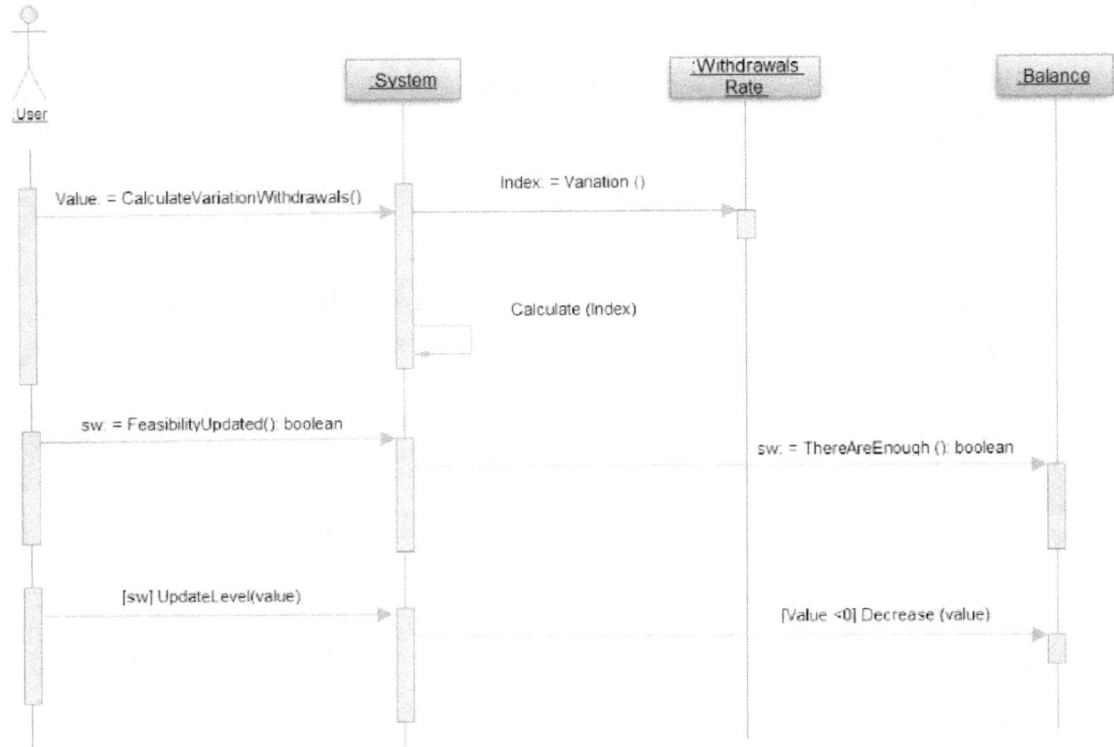

**Figure 50. Sequence diagram of the design phase of the withdrawals use case**

## Class diagram of the design phase

### *Introduction*

The Class Diagram of the design phase, describes graphically the specifications of the software classes, and interfaces in an application. The design class diagram usually contains:[1]

- Classes, associations and attributes
- Interfaces
- Methods
- Navigating
- Dependencies

---

[1] Larman. 1999

In Figure 51 is described as an example a class diagram, which specifies the format, in a first step, the attributes and methods whether public or private, in the second instance, associations, navigability and multiplicity; and finally the inheritance or generalization.

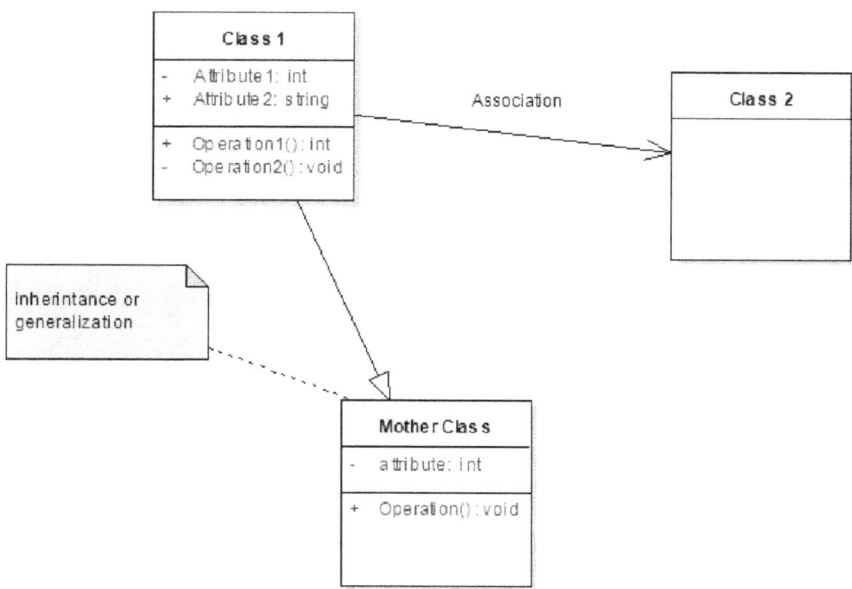

**Figure 51. Example of a Class Diagram of the design phase**


***Construction of Class Diagrams of the design phase, from the dynamics of systems under object orientation***

It can be easily conclude, by to compare the Forrester diagrams with the Conceptual Model, what then UML classes correspond to the Level subsystems and Flow subsystems of the Forrester Diagram. On the other hand, depending on their complexity, may create classes from other subsystems of Forrester diagram, as are the subsystems Auxiliary Source, Well, Random, etc. Now, we describe the most important features of classes based on flows and levels.

*Classes built based on the level subsystems.* These have one or more attributes(amount 1, amount 2, ...) in charge of preserving the storage level. Contain three methods: FactibilidadActualizar, ActualizarNivel and ObtenerNivelActual, where the first perform the function to check whether an update can be done, while the second makes the upgrade of storage, and the latter returns the current value stored in the level.

*Classes built on base of the flow subsystem.* These classes contain two attributes, Max and Min, indicating the limits of variation. You will also have one or more attributes of level type at which it is associated. Also have a method, CalculateVariation in charge of finding the value that will modify in the level class. Figure 52 shows an example of a class diagram of the design phase, which describes the classes built from the level subsystems and the Flow subsystems, whether input and output. By way of illustration shows the class diagram of the design phase of the problem of savings account in Figure 53.

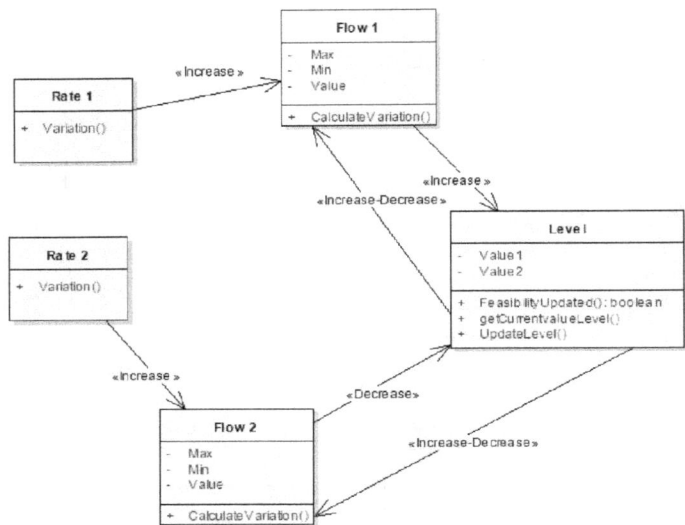

**Figure 52. General Class Diagram of the Design phase**

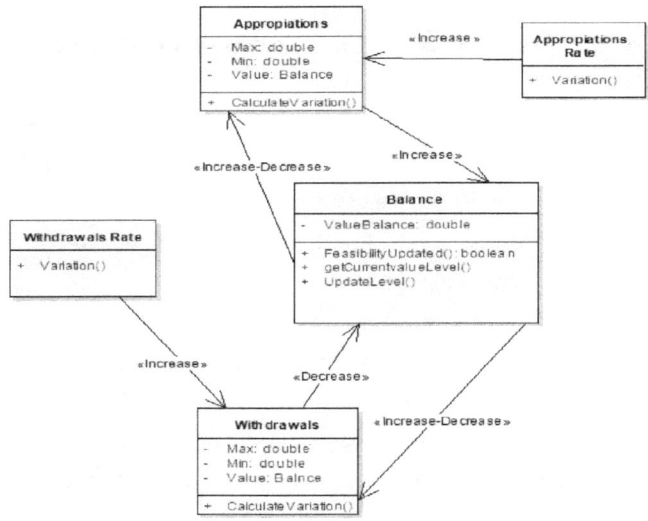

**Figure 53. Class diagram of the design phase of the case of savings account**

## Construction Sequence Diagram of the design phase, from the dynamics of systems, under the guidance of concurrent programming

Under the concurrent programming approach must take into account the flow subsystems represent the threads. Therefore in the Design Class Diagram should be added the inheritance of classes built from these subsystems with a thread object. It then describes the characteristics of the classes based on flows and levels in concurrent programming:

*Classes built based on the level subsystems.* These classes have the same characteristics as those built without the concurrent approach. Will charge the storage attributes and methods Level: FeasibilityUpdated, UpdateLevel and getCurrentValueLevel with the functions described in the previous section.

*Classes built on the basis of Flow subsystems.* First of all these classes contain the same two attributes(Max and Min), indicating the limits of variation, and those of type level to which it is associated, described for non-concurrent flow classes. As for his methods will CalcularVariacion and implementation of the method "runable" of wires that run call. The run() method is responsible for implementing the concurrent action of the class. In Figure 54 is described, in general, a Design Class Diagram in concurrent programming, and so in Figure 55 the case of savings account.

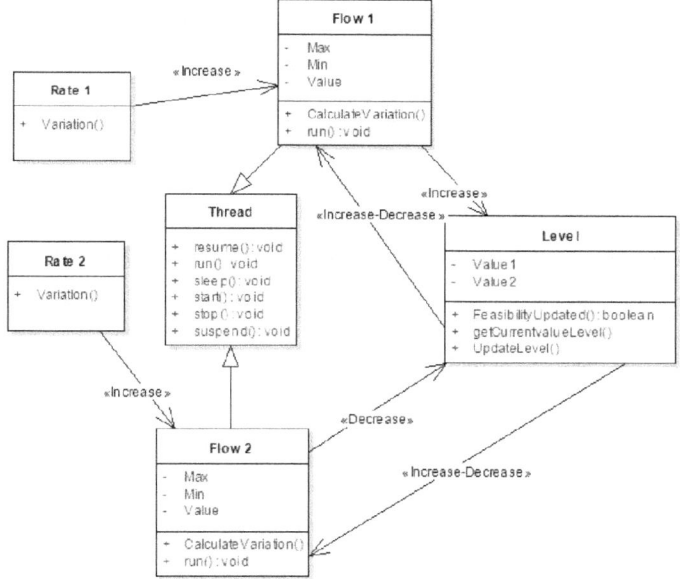

**Figure 54. Class diagram of the Design phase, Concurrent Programming based in Flow and Level Subsystems**

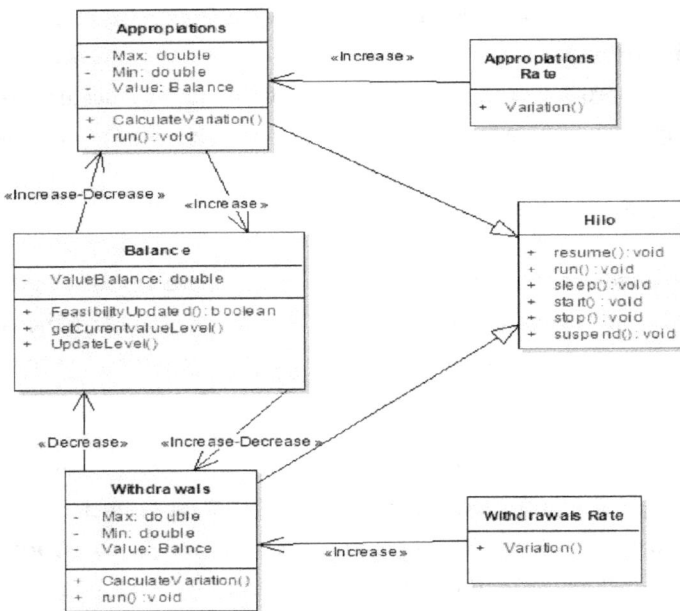

**Figure 55. Class diagram of the design phase in Concurrent Programming in case of savings account**

***Sequence Diagrams Building the design phase, from the dynamics of systems, under the guidance of Client Server Programming***

In the orientation of client server programming flow subsystems, and subsystem level, each generating a separate program. Levels generated server applications, and the flows, generate client applications. The following sections describe the characteristics of these applications.

*Application Servers built based on the level subsystems.* Servers present a class called VariableCompatida whose function is the treatment level storage. Also present a class of type thread, MainServer in charge for managing communication with customers, generating a ServerThread class instance for each client connecting.

The type thread class ServerThread is responsible for meeting all requirements of a particular Client and finally presents a user interface class whose mission, as well as communicating with the user, simply start and stop the main server.

*Client applications built based on the flow subsystems.* As in the application servers will SharedVariable class with similar function. Will also be a class of type thread called client charge of communication with the server and the associated variation in levels, and finally a kind of user interface. In Figure 56 is described, in general, a Class Diagram Design in concurrent programming, and so in Figure 57 the case of savings account.

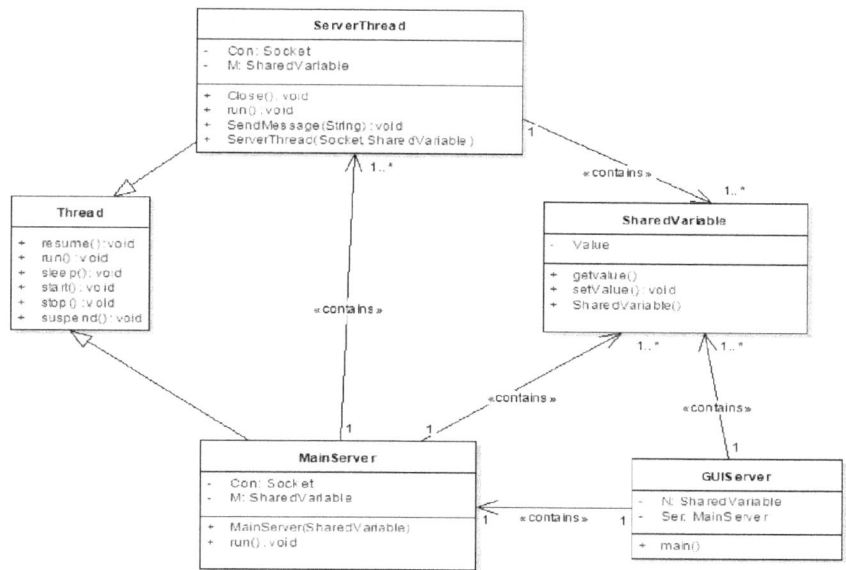

Figure 56. Class diagram of the design phase of a general server, based on level subsystems.

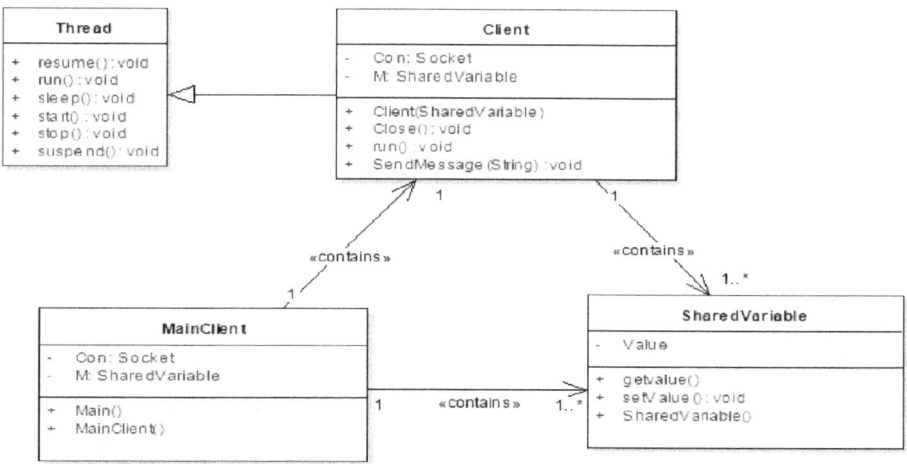

Figure 57. Class diagram of the design phase of a Customer General Flow-based Subsystems.

124 General Systems Theory: A focus on computer science engineering

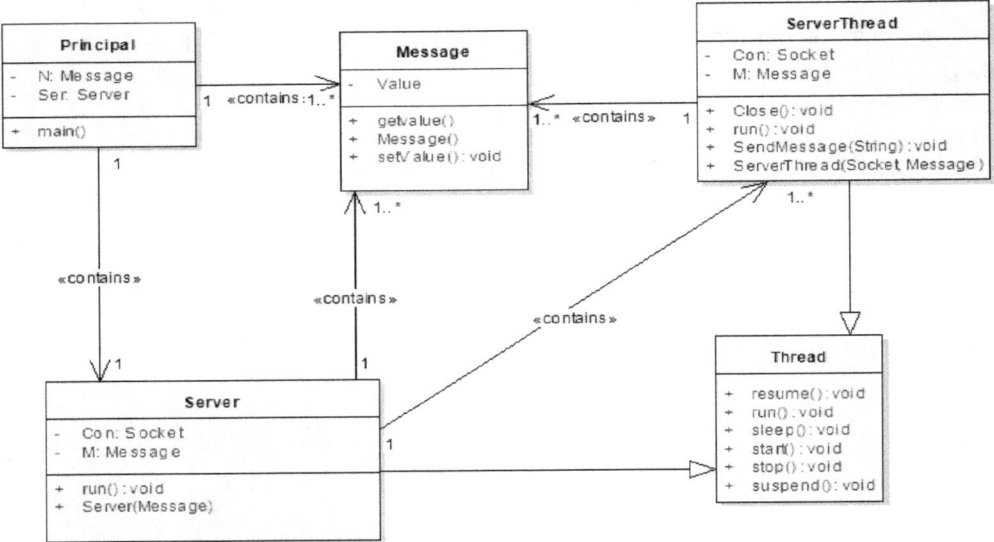

**Figure 58. Class diagram of the design phase, the server, the problem of savings account**

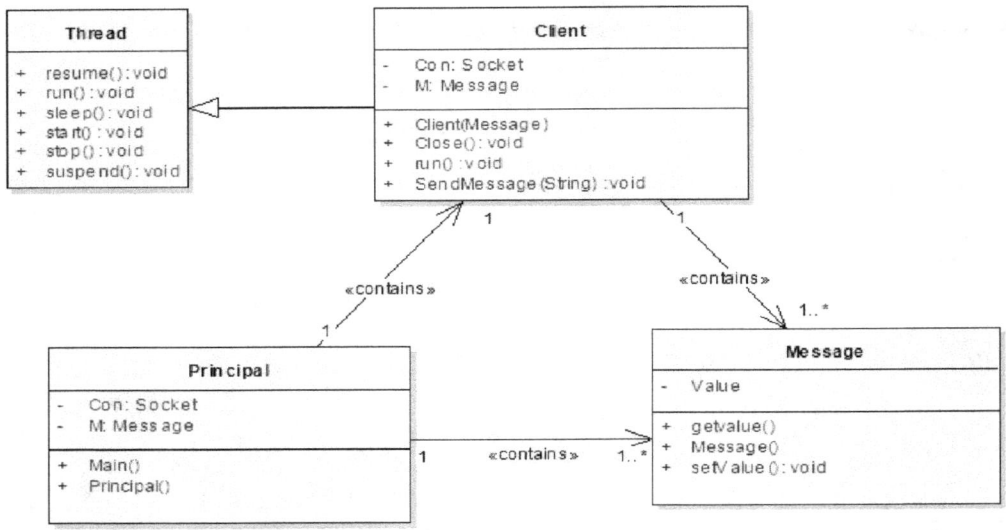

**59. Class diagram of the design phase, clients Appropriations, retreats and MostarSaldo, the problem of savings account**

# BIBLIOGRAPHY

*ARACIL, 1986*           ARACIL, Javier. Introducción a la Dinámica de Sistemas. Editorial Alianza. Madrid. 1986.

*ARACIL, 1997*           ARACIL Javier y GORDILLO Francisco. Dinámica de sistemas. Alianza Editorial, Madrid, 1997.

*ACKOFF, 1998*           ACKOFF, Russell L. El Arte de Resolver problemas, Noriega Editores- Decimocuarta reimpresión.- 1998

*BOOCH, 1999.*           BOOCH, G. Rumbaugh, J. Jacobson, I. El Lenguaje Unificado de Modelado. Addison Wesley. 1999.

*BERTALANFY, 001*        BERTALANFY, Ludwig von. Tendencias de la Teoría General de Sistemas. Alianza Editorial.

*BERTALANFFY, 002*       BERTALANFFY, Ludwig Von. Teoría General de Sistemas. Fundamentos, desarrollo, aplicaciones. México, Fondo Cultural Económica.

*CARRETERO, 2001.*       CARRETERO, Jesús et al. Sistemas Operativos: Una visión aplicada. McGraw-Hill. Madrid. 2001. 731 p.

*CHURCHMAN, 1973*        CHURCHMAN, West. El enfoque de sistemas. Editorial Diana. México. 1973

*COHOON, 2000*           COHOON, James, DAVIDSON, Jack. Programación y diseño en C++: Introducción al diseño y a la programación orientada a objetos. 2° Ed. McGraw-Hill.

*DEITEL, y DEITEL, 1999* DEITEL, H. M. DEITEL, P. J. Cómo programar en Java. 3° Ed. Prentice-Hall. 1999.

*HURTADO y NEIRA, 1996*  HURTADO, Dougglas. NEIRA, Marlon. Software aplicativo a la enseñanza de la asignatura sistemas operacionales. Tesis de Grado. Universidad del Norte. Barranquilla. 1996.

*JAWORSKI, 1999*         JAWORSKI, Jaime. Java 1.2 al descubierto. Prentice–hall Madrid. 1999. 1344 p.

*JOHANSEN, 1996*         JOHANSEN B, Oscar. Introducción a la teoría general de sistemas, – Decimotercera reimpresión - Noriega Editores, 1996.

*JOYANES, 2000*          JOYANES, Luis. ZAHONERO, Ignacio. Programación en Java 2:

| | |
|---|---|
| | Algoritmos, estructuras de datos y programación orientada a objetos. McGraw-Hill. Madrid. 2002. 725 p |
| **JOYANES, 1998** | **JOYANES, Luis.** Programación Orientada a Objetos. 2° Ed. Osborne McGraw-Hill. 1998. 895 p |
| **KENDALL, 1997** | **KENDALL, Kenneth. KENDALL, Julie.** Análisis y Diseño de Sistema. Pentice-Hall. 1997. 913 p |
| **LARMAN, 1999** | **LARMAN, Craig.** UML Y patrones: Introducción al análisis y diseño orientado a Objetos. Prentice hall. México.1999. 536 P |
| **LARMAN, 2002** | **LARMAN, Craig.** Applying UML and Patterns: An Introduction to Object-Oriented Analysis and Design and the Unifiqued Process. 2 Ed. Prentice Hall PTR. 2002. 627 P |
| **LATORRE, 1996** | **LATORRE, Emilio.** Teoría General de Sistemas. Aplicada la solución integral de problema. Editorial Universidad del Valle. Cali, 1996. |
| **LEA, 2001** | **LEA, Doug.** Programación concurrente en Java: Principios y patrones de diseño. 2° Ed. Addison Wesley. 2001. 430 p |
| **LEÓN-GARCÍA, 2002** | **LEÓN-GARCÍA, Alberto. WIDJAJA, Indra.** Redes de comunicaciones: Conceptos fundamentales y arquitecturas básicas. McGraw-Hill. Madrid 2002. 772 p. |
| **MAIN y SAVITCH, 2001** | **MAIN, Michael. SAVITCH, P.** Data structures and other objects using C++. Addison Wesley. 2001. 783 p |
| **PRESSMAN, 1998** | **PRESSMAN, Roger.** Ingeniería del software: un enfoque práctico. 4 ed. México: McGraw-Hill. Madrid. 1998. 581 p. |
| **ORTALI, 1998** | **ORTALI, Robert. HARKEY Don. JERI, Eduardo.** Cliente / Servidor: Guía de supervivencia. 2ed.Mexico: Mcgraw-Hill. 1998 |
| **SÁNCHEZ, 2001** | **SÁNCHEZ, Jesús et al.** Java 2: Iniciación y referencia. McGraw-Hill. 2001 |
| **SENN, 1992** | **SENN, James.** Análisis y diseño de sistemas de información. 2° ed. McGraw-Hill. México. 1992. 914 p |
| **SCHIDT, 1995** | **SCHIDT, Herbet.** C++ Manual de Referencia. Osborne McGraw-Hill. Madrid.1995. 592p |
| **TANAMBAUM, 1997** | **TANAMBAUM, Adrews,** Redes de computadores. 3° Ed. Prentice-Hall. Mexico. 1997. 812 p. |
| **TANAMBAUM, 1992** | **TANAMBAUM, Adrews,** Sistemas Operativos Modernos. Prentice- |

Hall. Mexico. 1992.

***TRUJILLO, 1996*** **TRUJILLO, Carlos**. Análisis de Sistemas. Editorial Universidad del Valle. Cali, 1996

www.ingramcontent.com/pod-product-compliance
Lightning Source LLC
Chambersburg PA
CBHW080917170526
45158CB00008B/2141